THE GENUS
ROSCOEA

A BOTANICAL MAGAZINE MONOGRAPH

THE GENUS
ROSCOEA

Jill Cowley

with contributions from

Richard Wilford

and

Roland Bream

Edited by J. M. Lock

Published by
The Royal Botanic Gardens, Kew

PLANTS PEOPLE
POSSIBILITIES

Text © The Board of Trustees of the Royal Botanic Gardens, Kew 2007
Illustrations © the artists, as stated in captions
Photographs © the photographers, as stated in captions
Jill Cowley has asserted her rights to be identified as the author of this work in accordance with the Copyright, Designs and Patents Act 1988

All rights reserved. No part of this publication may be reproduced, stored in a retrieval system, or transmitted, in any form, or by any means, electronic, mechanical, photocopying, recording or otherwise, without written permission of the publisher unless in accordance with the provisions of the Copyright Designs and Patents Act 1988.

Great care has been taken to maintain the accuracy of the information contained in this work. However, neither the publisher, the editors nor authors can be held responsible for any consequences arising from use of the information contained herein.

First published in 2007 by
Royal Botanic Gardens, Kew
Richmond, Surrey, TW9 3AB, UK
www.kew.org

ISBN 978-1-84246-134-1

British Library Cataloguing in Publication Data
A catalogue record for this book is available from the British Library

Production Editor: Ruth Linklater
Typesetting and page layout: Christine Beard
Design by Media Resources
Royal Botanic Gardens, Kew

Cover illustration: *Roscoea auriculata* painted by Christabel King. Plant cultivated in the Alpine Department, Royal Botanic Gardens, Kew.

Frontispiece: *Roscoea purpurea* forma *purpurea* painted by Christabel King.

Printed in Italy by Printer Trento

For information or to purchase all Kew titles please visit www.kewbooks.com or email publishing@kew.org

All proceeds go to support Kew's work in saving the world's plants for life

CONTENTS

LIST OF PAINTINGS vii

1. INTRODUCTION 1
   Acknowledgements 4

2. HISTORY OF THE GENUS ROSCOEA 7

3. WILLIAM ROSCOE (1753–1831) 11

4. MORPHOLOGY 17
   The Ginger family (Zingiberaceae) 17
   Description of the genus *Roscoea* 18

5. CYTOLOGY, ANATOMY, PALYNOLOGY & PHYLOGENETIC STUDIES 23
   Cytology 23
   Anatomy 23
   Palynology 24
   Phylogeny 25

6. ECOLOGY AND CONSERVATION 27
   Western Himalaya 28
   The Tibetan marches — Sichuan and Yunnan 29
   The North Burma Triangle 30
   Outlying species — Khasi and Chin Hills 31
   Substrate preferences 32
   Conservation 32
   Ecology in relation to cultivation 36

7. KEY TO THE SPECIES OF ROSCOEA 37

8. TAXONOMY OF SPECIES 39

9. HYBRIDS AND CULTIVARS **with contributions by Roland Bream** 153

CONTENTS

10. CULTIVATION by Richard Wilford 167
    Roscoeas in the open garden 167
    Growing Roscoeas under glass 172
    Propagation 174
    Pests and diseases 175

LIST OF EXSICCATAE 177

BIBLIOGRAPHY & REFERENCES 181

INDEX OF SCIENTIFIC NAMES 189

NEW ROSCOEA TAXA DESCRIBED IN THIS VOLUME

Roscoea alpina Royle forma **pallida** Cowley **forma nov.** 45
Roscoea cautleyoides Gagnep. var. **cautleyoides** forma **atropurpurea**
    Cowley **forma nov.** 117
Roscoea cautleyoides Gagnep. var. **cautleyoides** forma **sinopurpurea**
    (Stapf) Cowley **forma nov.** 117
Roscoea forrestii Cowley forma **purpurea** Cowley **forma nov.** 135
Roscoea humeana Balf. f. & W.W. Sm. forma **alba** Cowley **forma nov.** 129
Roscoea purpurea Sm. forma **alba** Cowley **forma nov.** 55
Roscoea purpurea Sm. forma **rubra** Cowley **forma nov.** 55
Roscoea scillifolia (Gagnep.) Cowley forma **atropurpurea** Cowley **forma nov.** 106
Roscoea tibetica Batalin forma **alba** Cowley **forma nov.** 140
Roscoea tibetica Batalin forma **albo-purpurea** Cowley **forma nov.** 141
Roscoea tibetica Batalin forma **atropurpurea** Cowley **forma nov.** 141
Roscoea tibetica Batalin forma **rosea** Cowley **forma nov.** 142
Roscoea tibetica Batalin forma **roseo-purpurea** Cowley **forma nov.** 141

LIST OF PAINTINGS

Plate 1. *Roscoea alpina* forma *pallida*; forma *alpina*. . . . . MARK FOTHERGILL (p. 40)

Plate 2. *Roscoea purpurea* forma *purpurea* . . . . CHRISTABEL KING (p. 46)

Plate 3. *Roscoea purpurea* forma *rubra* . . . . CHRISTABEL KING (p. 53)

Plate 4. *Roscoea capitata* . . . . CHRISTABEL KING (p. 57)

Plate 5. *Roscoea ganeshensis* . . . . CHRISTABEL KING (p. 61)

Plate 6. *Roscoea tumjensis* . . . . CHRISTABEL KING (p. 68)

Plate 7. *Roscoea auriculata* with habit sketch . . . . CHRISTABEL KING (p. 74)

Plate 8. *Roscoea brandisii* (as *Roscoea purpurea*). . . . . WALTER HOOD FITCH (p. 84)

Plate 9. *Roscoea australis*. . . . . CHRISTABEL KING (p. 87)

Plate 10. *Roscoea wardii* . . . . CHRISTABEL KING (p. 92)

Plate 11. *Roscoea schneideriana* . . . . MARK FOTHERGILL (p. 97)

Plate 12. *Roscoea scillifolia* forma *scillifolia*; forma *atropurpurea* . . . CHRISTABEL KING (p. 102)

Plate 13. Roscoea *cautleyoides* var. *cautleyoides* forma *cautleyoides*
. . . . WILLIAM EDWARD TREVITHICK (p. 108)

Plate 14. *Roscoea cautleyoides* var. *cautleyoides* forma *sinopurpurea* . . CHRISTABEL KING (p. 113)

Plate 15. *Roscoea humeana* forma *humeana*. . . . . WILLIAM EDWARD TREVITHICK (p. 119)

Plate 16. *Roscoea humeana* forma *lutea* . . . . CHRISTABEL KING (p. 125)

Plate 17. *Roscoea forrestii* forma *forrestii* . . . . CHRISTABEL KING (p. 132)

Plate 18. *Roscoea tibetica* forma *alba*; forma *albo-purpurea*;
forma *roseo-purpurea* and forma *rosea*. . . . . CHRISTABEL KING (p. 136)

Plate 19. *Roscoea praecox* . . . . CHRISTABEL KING (p. 146)

Plate 20. *Roscoea* × *beesiana* . . . . CHRISTABEL KING (p. 157)

In memory of Rosemary Smith (1933–2004)

who first drew my attention to *Roscoea*, and who helped me
to understand the intricacies of the ginger family.

1. INTRODUCTION

"There are two or three sorts of Roscoea *— there's a yellow one............. and a pink one and there's a blue one."*
O.A.P. Wyatt VMH
(Journal of the Royal Horticultural Society 92(6): 247 (1967)

If only it were this simple!!

In the last few years, interest in *Roscoea* among growers and gardeners has grown apace. This book aims to provide correct names for those species already in cultivation, as well as discussing those species that may form exciting additions to our gardens in the future.

Over sixty years ago, J.M. Cowan of the Royal Botanic Garden, Edinburgh was querying the identification of plants grown there, in his review of the genus (1938). However, some of the names used in his paper and elsewhere were wrong, and these incorrect names continued in use because there were no more recent detailed studies. I began to study the genus in 1978, and published a revision in 1982 that addresses most of the problems of which I was aware at that time. The correct names of wild species and of plants in cultivation are, I believe, now reasonably well established. In addition, several new species and varieties have been described as new to science since my 1982 revision.

It is no wonder that confusion has arisen and misidentifications have occurred, because the genus includes many species that are incredibly variable in form and colour. For instance, the appearance of the plants may change considerably with age. When the first flowers emerge, sometimes without the leaves, the plants can look completely different to mature specimens that one sees weeks or months later. Many species flower over a long period although, as Roy Lancaster pointed out in his excellent article on *Roscoea purpurea* (1992), these are likely to be the less spectacular species.

Any student of the *Zingiberaceae* knows that dried specimens are difficult to identify. Preserving flowers for future identification is not easy; *Roscoea* is no exception. The flowers usually last only a single day. The perianth segments are very thin and become stuck together when pressed, so that examining the separate parts later is almost impossible. The only way around this problem is to carry spirit or alcohol into the field and to preserve the flowers in it. Cowan (1938) wrote of *Roscoea*: "The species are difficult to delineate in this genus, the criteria used to distinguish them are quite unreliable; further, roscoeas do not dry well; short of entirely destroying specimens it may be impossible to name them; the original type specimens are often poor and descriptions brief. Our knowledge of certain species is therefore quite inadequate, so that to name roscoeas and to assign the many recently found intermediate forms to their proper places is no straightforward task". When I was working on my revision (Cowley 1982), study of fresh flowers of the few species in cultivation greatly helped me to interpret dried material.

Other difficulties in the taxonomy of the genus arise because the plants change their appearance from seedling to maturity, as well as from the first flowering at the beginning of the season to the fully developed fruiting plant. Some species have compact forms with non-exserted peduncles, and others with taller exserted peduncles in the same population; when seen in isolation, these may look like two distinct species. Habitat can also influence the development of plants in a population; those growing through other vegetation appear different from those in the open. I saw differences of this kind when I studied *Roscoea humeana* and *R. scillifolia* in cultivation, and then later when I saw *R. humeana* and *R. tibetica* in the wild. As the majority of the herbarium specimens show, plants of *R. humeana* growing in rock crevices were robust, with larger flowers and immature leaves at flowering time. However, plants growing in the open, in cultivation, tend not to flower until the leaves are well developed, and the whole plant appears less robust. In *R. scillifolia*, which unfortunately I have not seen in the wild, plants growing under adverse conditions, such as through other vegetation, had longer leaves and inflorescences with exserted peduncles. *Roscoea tibetica* showed the most variation in the wild; the collections that I made from different habitats maintained their differences in cultivation. There is more on this topic in the individual species accounts.

It must also be said that plants that have been in cultivation for a long time sometimes seem to take on a different appearance, and can be difficult to correlate with wild collections. Also, although I have seen no evidence of hybridisation in the wild, it may occur in cultivation.

Most Botanic Gardens now prefer to grow only plants of known wild provenance. These have greater scientific value than those of unrecorded origin. In the past, it was enough to possess a particular species, regardless of source. For aesthetic purposes this may be reasonable, but for research projects such plants have limited use, and only in cases when wild-collected and well-documented specimens are impossible to obtain is there a rationale for maintaining plants of unknown origin.

Valuable information that comes with the plant to the nursery or garden is often not regarded as important, or becomes lost over the years. We now know that it is essential to retain collector's numbers, and any other information, if the plant is to retain any scientific value. As an instance of this, Gary Dunlop sent me a photograph of a *Roscoea* for naming, and I noticed that it had a label attached, with a number. I recognised the number as being that of the Kingdon Ward collection which I had used as the type of *Roscoea australis*. The scant material that I saw when I described this new species consisted of a herbarium specimen and a flower in spirit from the Royal Botanic Garden, Edinburgh. This showed that the species had once been in cultivation. Kingdon Ward always sent his seeds and plants to gardens and interested nurserymen, and Gary Dunlop's plant showed that the species was still in cultivation. The illustration of this species in this book (Plate 9) was prepared from a plant generously sent to Kew by Mr Dunlop.

We owe a huge debt of gratitude to intrepid plant hunters, like Kingdon Ward, who for months (and sometimes years) at a time worked tirelessly in difficult conditions, recording and collecting herbarium specimens and trying to keep plants and seeds viable. Communication was often difficult, and they had to contend with difficult terrain, appalling weather, lack of transport, and the suspicions of the local people. They suffered enormous hardships so that we could enrich our gardens and greatly expanded our knowledge of the unique and diverse flora of many regions.

The main collectors of *Roscoea*, or those who had plants collected for them, in the Himalaya, included Dr Nathaniel Wallich (1786–1854), Dr John Forbes Royle (1798–1858), Sir Joseph Dalton Hooker (1817–1911), Frank Kingdon Ward (1885–1958), Frank Ludlow (1885–1972), and Major George Sherriff (1898–1967) and more recently the trio of Oleg Polunin, Sykes and Williams. Those who collected *Roscoea* in southwest China and Tibet included Jean Marie Delavay (1834–1895), Augustine Henry (1857–1930), George Forrest (1873–1932), Camillo Karl Schneider (1876–1951), Heinrich von Handel-Mazzetti (1882–1940), Joseph Francis Charles Rock

(1884–1962) and again Frank Kingdon Ward (1885–1958). It must not be forgotten that collectors such as Forrest (McLean 2004) and Kingdon Ward were sponsored by Arthur Kilpin Bulley (1861–1942), who founded the nursery firm of Bees Ltd. of Liverpool (McLean 1997). Another nurseryman who sponsored collectors in this way was Sir Harry James Veitch (1840–1924). He financed the plant hunter Ernest Wilson to travel to Sichuan in 1899; this expedition was followed by later ones. Rich gardeners who wanted the latest plant discoveries for their estates also financed plant hunters at the beginning of the last century.

When I began my revision of *Roscoea*, much of the cultivated stock at Kew and elsewhere was of unrecorded origin. The (Zhongdian) Chungtien-Lijiang-Dali Expedition to Yunnan in 1990 (CLDX) and the Baker, Burkitt, Miller & Shrestha (BBMS) Expedition to Ganesh Himal, Nepal in 1992 changed the situation dramatically at Kew, and subsequently the collection has been much enhanced by further new introductions.

I was lucky enough to be a member of the 1990 CLD Expedition to Yunnan province of China, of a visit to Yunnan and Sichuan led by Philip Cribb in 1992, and of the ACE (Alpine Garden Society China Expedition) in 1994. These visits allowed me to see species growing in populations in their native habitat, and to make observations which would have been impossible on plants in cultivation. I reintroduced *Roscoea schneideriana* to cultivation in 1990, and introduced *R. praecox* in 1994. The excitement of finding the latter species by the roadside in red soil, where earlier writers had described it as growing, made some of the difficulties of travelling in China worthwhile. Such journeys have their delights and surprises all the time; you never know what is going to be around the next bend, what the plumbing will be like in the next inn, or what your next meal will consist of. At first, I was a little reticent about eating *Lilium davidii* bulbs, as I feared they were an endangered species, but when in Rome and all that…..: they were delicious (nutty) as were the grasshoppers (crunchy). Yak-butter tea and doughbuns were a different story.

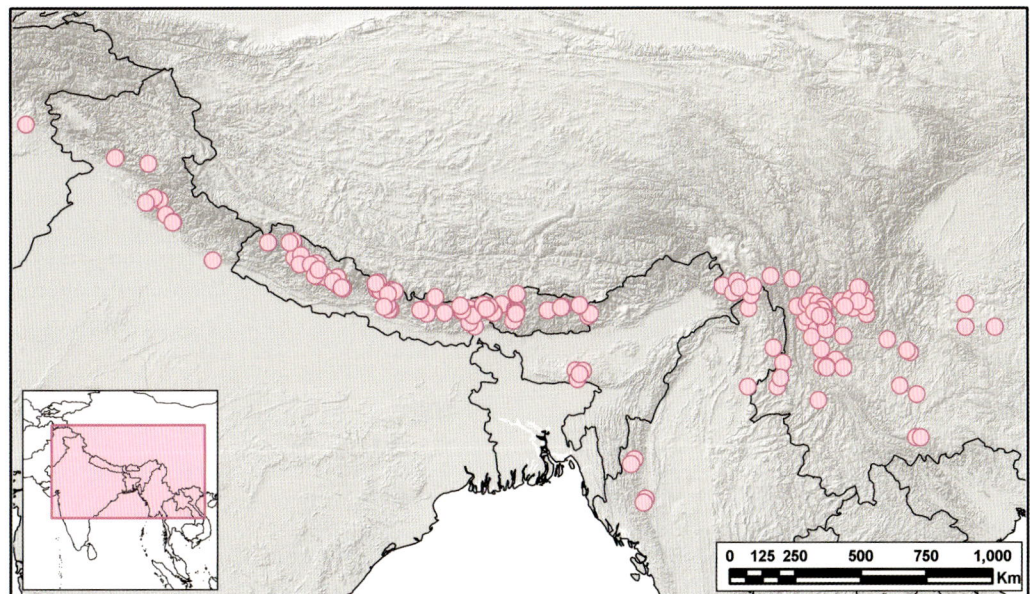

Map 1. Range of distribution of the Himalayan group and the Chinese group of species.

It was a treat to see *Roscoea humeana* in small drifts across the Lijiang Plain in Yunnan, in all the pale colours: white, pale purple and bi-coloured, not the familiar purple ones in cultivation in Europe (Figs. 69, 70). We saw populations of both the purple and yellow *R. cautleyoides*, but the two colours were seldom in the same place; there was just the odd yellow-flowered plant in a purple population and vice versa. There were no intermediates anywhere. Further up the valley from the Plain, the Gang Ho Ba (a dried-up river-valley) was not to disappoint. There were drifts of the yellow *R. cautleyoides* (Figs. 58, 66) and *R. humeana*, and in the scree areas, *R. tibetica*.

In the old days it was easy to share the material collected by the expeditions. Today it is not so easy. There are very strict laws in place to prevent people removing genetic resources, for commercial gain, from other countries without the informed consent of the country involved. The United Kingdom has ratified the proposals put to the Biodiversity Convention at the Rio Summit held in 1992. Scientific institutions, like the layman, must obtain the correct permissions in order to undertake scientific research in other countries which have ratified the Agreement. Plant material collected for scientific purposes cannot be transferred to commerce where revenue would be made from a country's genetic resources.

Locations mentioned in the text have been copied as written by the collector. A comprehensive glossary of Chinese place names can be found in Lancaster (1989: 478–481). Another can be found in Winstanley's (1996: 180) English translation of Handel-Mazzetti (1927).

It is possible to have hardy species of *Roscoea* in flower in succession from April until October by choosing species from the range now in cultivation. More information can be found in the chapters on 'Cultivation' and 'Cultivars and hybrids'.

Most species appear to be endemic to small vulnerable areas and this monograph can only cover the known species. Hopefully more new species will be found in the future. I feel sure that there are many new species in the unexplored river valleys of the Himalaya.

ACKNOWLEDGEMENTS

My thanks go to Wessel Marais and Brian Mathew, my Kew colleagues in the Petaloid Monocotyledon Section, for their patience, friendship and encouragement and for giving me the confidence to develop my *Zingiberaceae* studies.

Thanks to Rosemary Smith (formerly of Edinburgh Botanic Garden) for her friendship and for all her careful instruction and suggestions.

My thanks go to Chris Brickell for giving me the opportunity to see *Roscoea* in the wild on the CLDX (1990) and ACE (1994) expeditions.

I am very grateful to Professor Li and Mrs Zhang Changqin of the Kunming Institute for their friendship, companionship and guidance on my first daunting expedition to China in 1990.

Thanks to Bill Baker for his diligence and interest in finding so many species on his Ganesh Himal expedition in 1992.

Grateful thanks also go to Tony Hall, Graham Walters, Richard Wilford and other staff members of the Alpine Department at Kew for their interest in the successful cultivation of many *Roscoea* species new to cultivation; without their skills there would be no illustrations.

The beautiful paintings for the plates and the dissection figures were executed by Christabel King and Mark Fothergill with their usual skill, for which I am truly grateful.

I am also grateful to the Librarian at Edinburgh Botanic Garden, Mrs Jane Hutcheon, for allowing me to use one of Oleg Polunin's photographs to illustrate *Roscoea nepalensis*.

Thanks go to Justin Moat and his GIS department for the distribution maps.

The publications by Dr Chatchai Ngamriabsakul from Thailand and Dr Mark Newman from the herbarium at the Royal Botanic Garden Edinburgh, have added greatly to the content of this book, for which I thank them.

I am indebted to Richard Wilford for writing the cultivation chapter so expertly.

Valuable correspondence concerning cultivation of *Roscoea* was exchanged with Gary Dunlop, Trevor Healey and Liz White. To Roland Bream, who holds the NCCPG *Roscoea* collection, many thanks for his hospitality and unwavering enthusiasm for the genus. Thanks are also due to him for the notes on hybrids and cultivars.

Thanks go to Melanie Thomas for invaluable help with the latin diagnoses.

Last but by no means least, my very grateful thanks to my brother, Paul Bedwell, for saving me from too many panic attacks when the computer got the better of me.

To all those lovers of *Roscoea* who have been waiting patiently for this book to appear, I apologise for the long delay.

2. HISTORY OF THE GENUS ROSCOEA

Plants that are now assigned to the genus *Roscoea* first became known to European botanists from material collected by Dr Buchanan at Narainhetty, a Hindu temple opposite the Palace in Kathmandu, Nepal and now called Narayan Hiti (see page 8). Some of the specimens are housed at the Linnean Society, London, as part of the Smith Herbarium. The specimens came to the attention of J.E. Smith, who described them as *Roscoea purpurea* in his *Exotic Botany* 2: 97 (1805–1808), and illustrated them in his Plate 108, executed in 1806, of which the original is housed in the Central Library in Liverpool. In 1807, William Roscoe, after whom the genus had been named, published his great work on the ginger and related families: *A new arrangement of the plants of the Monandrian Class usually called Scitamineae*.

During the period of British control in India, many collectors operated in the highland regions, where staff went during the hot season for rest and recuperation. One of the first of these was Nathaniel Wallich, who collected in Nepal in 1819–22, and among his collections was the specimen which was to become the type of *Roscoea capitata* (Smith 1822). A little later, Lady Dalhousie collected in Simla, in the western Himalayas. Simla was the seat of Government in India during the hot season, and many other collectors operated from there. John Forbes Royle, the author of *Illustrations of Himalayan Botany*, published *Roscoea alpina* in 1839.

Somewhat later, in 1845, Thomas Thomson collected around Mussourie, Indian Himalaya. In 1849 and 1850 Joseph Dalton Hooker, the son of the first Director of Kew, Sir Willliam Jackson Hooker, collected in Sikkim. Among his collections were the specimens from which *Roscoea auriculata* was described by K. Schumann. In 1850 Hooker & Thomson collected together in Assam.

A quarter of a century later (1872–1876), C.B. Clarke, later a major contributor to the *Flora of British India* and other Kew floras, collected in Assam, and later in Kashmir and, during 1885, was with his collectors in Sikkim. Another botanist, J.S. Gamble, known for his *Flora of Madras*, collected in Simla in 1877; in 1898 he was again collecting, but in the Himalayan area of Kumaon. In 1878–1879 the first known local collector of these plants, Dungboo (this is all we know of his name) collected in Tibet. In 1884–1887, Dr King and his collectors were in Sikkim and Assam, and during 1888, J.R. Drummond collected in Kashmir.

British botanists in India were mainly civil servants, but in China many of the early collections were made by French missionaries. The first of these to collect *Roscoea* seems to have been Jean Marie Delavay, who collected in Yunnan, China in 1883. Among his collections was the specimen which was to be described by Gagnepain, at the Paris herbarium, as *R. cautleyoides*. Later, between 1887 and 1889, he also collected the specimen which was to become the type of *R. scillifolia*. He returned to Yunnan in 1897. Russians were also travelling and collecting in China, and in 1893 Kachkarov collected the specimen that was to become the type of *R. tibetica*, from Sichuan. Another Frenchman, Augustine Henry, a Customs official, was stationed in Mengtsz, Yunnan, China, and collected in the area in 1896. One of his collections was to become the type of *R. praecox*. In 1898–9, François Ducloux collected in Yunnan. Two of his collections were to become the types of *R. intermedia* var. *anomala* and var. *macrorrhiza*, both now considered synonyms of *R. praecox*.

Collections of *Roscoea* were now becoming sufficiently numerous for taxonomists to begin to produce accounts of the genus as a whole. In 1902, François Gagnepain, who worked in the herbarium at Paris, described several species and varieties: *R. intermedia* (*R. alpina*), *R. intermedia* var. *minuta* (*R. tibetica*), *R. intermedia* var. *anomala* (*R. praecox*), *R. intermedia* var. *macrorrhiza* (*R. praecox*), *R. capitata* var. *purpurata* (*R. cautleyoides*), *R. capitata* var. *scillifolia* (*R. scillifolia*), *R. cautleyoides*, *R. debilis*, *R. chamaeleon* (*R. cautleyoides*). (The currently accepted names are placed in brackets). In Germany, Engler's monumental *Das Pflanzenreich* was slowly taking shape, and in 1904 K. Schumann published his account of the family *Zingiberaceae* within this. He described three new species (*R. praecox*, *R. auriculata* and *R. blanda*), and raised Baker's var. *brandisii* of *R. purpurea* to full specific rank as *R. brandisii*.

The number of collectors in Himalayan regions and in the mountainous regions of China now began to increase greatly. Haines collected in Mussourie, Indian Himalaya, in 1906, and Venning collected in Burma in 1910. An era of commercial plant collecting, financed by rich nurserymen and gardeners in Britain, Europe and America now began. George Forrest made his first expedition of several in 1904–1907, collecting in Yunnan. He was to return several times to Yunnan and Sichuan in 1910–1911, 1912–1914, 1917–1920, 1921–1923, 1924–1926 and 1930–32. The type specimen of *Roscoea forrestii* was collected in 1913. His first sponsor was the nurseryman A.K. Bulley, who also sent another collector, R.E. Cooper, to Sikkim in 1913 and then to Bhutan in 1914. In 1916, a plant grown at Edinburgh from one of George Forrest's earlier seed collections was described as *Roscoea humeana*, commemorating a young member of the staff of the Edinburgh Botanic Garden who died in the Great War. Forrest died in China in 1932.

While he was in Yunnan and Sichuan in 1914, Forrest met the Austrian botanists, Heinrich von Handel-Mazzetti and Camillo Schneider. The former travelled widely in the region, and in 1914 he collected the type of *Roscoea schneideriana* in Yunnan, and later, in 1916, in the under-collected area around the Yunnan/Tibet/Burma borders, the type of *R. blanda* var. *pumila*, now regarded as a synonym of *R. wardii*.

Roscoea wardii commemorates another prolific collector of plants now much grown in European gardens, Frank Kingdon Ward, who was collecting in Burma in 1914 and

Herbarium specimen from the original collection, made in 1802, by Dr Buchanan at Narainhetty, Kathmandu, Nepal, of *Roscoea purpura* Sm., the type of the genus *Roscoea*. Housed at World Museum Liverpool. Reproduced courtesy of the Board of Trustees of National Museums Liverpool.

again, in Yunnan and Sichuan, during 1921. He made several more collecting expeditions, visiting Burma and Assam in 1926 and returning to Assam in 1928–1929. In 1931 he was in Burma, and Tibet in 1933. In 1935–1938 he returned to Assam. After World War II he returned to Assam in 1949–1950, and on his last expedition to Burma in 1956, he collected the plant that was to be described as *R. australis*. Other collectors, including the English plantsman Reginald Farrer, collected around this time; Farrer collected on the Burma-Yunnan border in 1920. An American botanist, Joseph Rock, joined up with Handel-Mazzetti to collect in Sichuan in 1922.

These collections stimulated further taxonomic work; in Berlin, Th. Loesener published an article on various Chinese taxa of *Roscoea* from Yunnan, partly based on taxa described earlier by Gagnepain (1901/1902). Unfortunately Loeseuer omitted to say which he believed to be the typical variety of his species *R. yunnanensis* which is now referred to *R. cautleyoides*. Loesner treated the following taxa: *R. yunnanensis* (*R. cautleyoides*), *R. yunnanensis* var. *purpurata* (*R. cautleyoides*), *R. yunnanensis* var. *schneideriana* (*R. schneideriana*), *R. yunnanensis* var. *dielsiana* (*R. schneideriana*), *R. yunnanensis* var. *scillifolia* (*R. scillifolia*), *R. intermedia* var. *plurifolia* (*R. tibetica*), *R. blanda* var. *limprichtii* (*R. debilis* var. *limprichtii*). (Currently accepted names are in brackets).

Collectors continued to operate; Camillo Schneider collected in Sichuan in 1926, and Joseph Rock in Sichuan in 1928–1929. Rock returned and travelled in Sichuan and Yunnan in 1931–1932. A partnership whose names now became prominent was that of Frank Ludlow and Major George Sherriff, who collected in Bhutan in 1933. In 1936, they turned their attention to Tibet while Henry Duncan McLaren (2nd Lord Aberconway) had plants collected for him in Yunnan. In 1938, B. J. Gould was collecting in Bhutan and a year later he was in Tibet. During 1939, Sherriff, apparently alone, collected around Simla, western Himalayas.

1938 saw the first attempt at a horticultural review of the genus. This was made by J.M. Cowan, who was working in Edinburgh, where he would have seen the Forrest collections in the Herbarium, and also living plants at the Botanic Garden. This was an excellent review, bearing in mind the somewhat limited and conflicting information that was available at the time. For example, he rightly questioned the identification as the Himalayan *Roscoea alpina* of the plants grown in gardens from seed collected by Forrest in China. These have subsequently been identified as *R. scillifolia*. The review was accompanied by photographs of four species growing in the Edinburgh Botanic Garden at that time: *R. scillifolia*, *R. schneideriana*, *R. cautleyoides* and *R. humeana*. In the same year Charles Hervey Grey published an article on *Roscoea* in a work on *Hardy Bulbs*, based on herbarium collections alone.

After the Second World War, Kingdon Ward (see above) was the first collector back in the field, visiting Assam and Burma. Among his collections from Assam, was the specimen later to be described as *Roscoea wardii*, and on his penultimate expedition, to Burma, he discovered the plant that I have described as *R. australis*. In 1949–1950, Oleg Polunin collected in the Langtang valley and other parts of Nepal, while Ludlow, Sherriff and Hicks were in Bhutan. In 1962, Stainton was again in Nepal, this time visiting the Ganesh Himal region. In 1964 D. McCosh was in eastern Nepal and in 1965, A.D. Schilling, Sayers and others collected in the Langtang valley.

1970–73. Four more expeditions to Nepal took place during this period, the first, in 1970, by Brian Halliwell who brought back living material for Kew. Sir Colville Barclay and Patrick M. Synge were there in 1971, and Kanai, Hara & Ohba explored the Trisuli Khola in 1972. The last foray in this period was by Chris Grey-Wilson and Barry Phillips in 1973.

In the days before the closure of Bhutan to travellers and scientists, Douglas Grierson and David Long from Edinburgh Botanic Garden collected there in 1975, studying the flora for the projected *Flora of Bhutan*.

Not long after this I began to work on the genus, and in 1982 I published a revision of the genus, based on all the material available to me at that time. I described four new species: *Roscoea tumjensis*, *R. wardii*, *R. australis* and *R. forrestii*. I also made three new combinations: *R. schneideriana*, *R. scillifolia* and *R. debilis* var. *limprichtii*.

In 1990 I was lucky enough to be invited to represent Kew on the first part of the (Zhongdian) Chungtien-Lijiang-Dali Expedition (CLDX), which also included representatives from the Royal Horticultural Society and the Royal Botanic Garden, Edinburgh. The expedition visited China in June and October. Herbarium specimens and living collections were made but it was mainly a seed collecting trip. Among the welcome collections of *Roscoea* brought back for Kew was material which turned out to be *R. schneideriana* and allowed its reintroduction to cultivation.

The year 1992 saw the Oxford University Expedition to the Ganesh Himal in Nepal, mainly to study orchids although the participants, Bill Baker, Tom Burkitt and J.T. Miller succeeded in finding some very interesting plants. Kew benefited from donations of living material of all the species of *Roscoea* known from the area, accompanied by full locality data. These included a formerly unseen colour variant of *R. purpurea*, one clone of which was to be described as the cultivar 'Red Gurkha'. I later described one collection from this expedition as *R. ganeshensis*.

In 1994, representatives of the Royal Horticultural Society, the Royal Botanic Garden, Edinburgh and Kew, together with eminent nurserymen, came together in China for the Alpine Garden Society Expedition to China (ACE). Again the expedition had two parts, summer and autumn, and aimed mainly to collect seed. From this, five colour forms of *Roscoea tibetica* were brought into cultivation.

More recent work on the genus has led to the publication of *Roscoea kunmingensis* and its variety *elongatobractea* by S.Q. Tong in 1992. Recently the Thai botanist, Dr Chatchai Ngamriabsakul, working at Edinburgh, has studied the systematics of the genus, using new evidence from cytology and molecular studies. This work led to the description of a new species, *R. bhutanica*, in 2000.

3. WILLIAM ROSCOE (1753–1831)

"In the mountains, at 9,000 feet, in North Western Himalaya and Nepal, grow plants of a small distinguished group, the Roscoeas. They have raised their flowers to the eternal heights for countless ages; they will repeat their life-rhythms for — who knows! — long beyond the span of humans on Earth. The memorials of men to their illustrious kin, created in statuary or paintings will crumble and fade, the Roscoeas will persist, and each season, in freshness and brightness perpetuate the memory of a remarkable Liverpool gentleman — William Roscoe. H. Stansfield**

Roscoe was certainly a remarkable gentleman, coming as he did from humble beginnings; he was a self taught, self motivated person who in his time became a competent historian, poet, painter, collector, politician, writer, banker and by no means least, naturalist and plantsman. Just a couple of these accomplishments would satisfy most people and like his namesake William Morris, who was born three years after Roscoe's death, one muses that they must have had days with more hours in them than other mortals. By all accounts Roscoe was, unlike Morris, able to keep the attentions of his wife as well as indulge in all his other activities.

Roscoe was born on 8th March 1753 at the Old Bowling-Green House on Mount Pleasant, a turnpike road in Liverpool. Very soon after his birth the family moved not very far away to a place where there was again a bowling green and his father, also William, became a tavern owner. Close by, his father also ran a small market garden which was to stimulate young William's interest in agriculture and natural science.

At the age of six he came under the tuition of a Mr Martin at a school in Paradise Street in the centre of Liverpool. He obviously took to this teacher, later admitting that from him and his mother he learnt good humanitarian principles by which he was to live later in his life. Mr Sykes had the unenviable task of teaching him arithmetic, writing and English grammar. William was, however, rather unruly, having an aversion to compulsion and restraint and was happy to cease school attendance at the age of twelve. At this age he was self-willed, a wild child with an unsociable disposition, preferring to be alone in the open air, strolling by the Mersey, going for long walks or fishing. He had visions of being a sportsman, and managed to get hold of a gun and shot a thrush. He was horrified at what he had done and vowed never to do it again.

At this time he was happy helping his father in his market garden, carrying baskets of potatoes to the market on his head and generally caring for the garden. He considered that those people who cultivate the earth by their own hands must be the happiest. Next to his father's property was Reid's factory, manufacturing British chinaware. Transfer printing had just been invented and William became friendly with the painters and helped them, becoming "tolerably expert". One of the painters, Hugh Mulligan, was an engraver of copper plates and a poet, who published books of poems. He became William's mentor, starting a friendship which would continue until Mulligan's death. During this time William also tried his hand at joinery, making a bookcase which he filled

---

*Courtesy of the Board of Trustees of the National Museums & Galleries on Merseyside.

with volumes of Shakespeare whose plays he memorised. He was also at this still early age fascinated by poetry and was soon to be producing his own. In his hours away from the garden he would be an avid reader and at 15 chose to become a bookseller, working for a Mr Gore, a Liverpool tradesman. However, it seems that this did not suit him, and he left after a month.

In 1769 he became articled for six years to a Mr John Eyes Jnr., an attorney and solicitor. Roscoe's time outside work was spent in educating himself and in developing a love of the arts and literature, especially all things Italian. However work had to come first, as by this time his mother had died and his father and sister were both dependant on William. He was determined to acquire as much knowledge as he could about his profession so that he could be useful to his employer. After Mr Eyes's death William's clerkship came under Mr Peter Ellames, an eminent Liverpool attorney, who was pleased to have such a conscientious worker as Roscoe in his employ.

At this time William was acquiring friends with the same taste in studies, in literature and in the beauties of nature. All through his life he attracted people to his warmth, open-heartedness, generosity, his versatile intelligence and his enthusiasm for what life had to offer; he had few enemies even when later he stood for principles which the majority, at the time, scorned. One of his early friends was Francis Holden who had an uncle who was a linguist, and so it was that Roscoe started to study languages. Holden was to move away from Liverpool to Glasgow and then to France, but the friendship remained intact until Holden's return to Liverpool. By this time Roscoe had made friends with Mr William Clarke and Mr Richard Lowndes with whom he studied dead languages, and on Holden's return all four intellectual friends pursued their classical studies. William Clarke used to travel to Italy and bring back books for Roscoe.

In 1773, Roscoe became one of the founders of the Liverpool Society for the Encouragement of the Arts of Painting and Design, and in 1774 the Society mounted its first art exhibition. Much later Roscoe was to become known as one of Liverpool's greatest patrons of the Arts.

In 1774 Roscoe was admitted as an attorney of the Court of King's Bench and practiced in Liverpool, working with Mr Samuel Aspinall. Later he became a partner in the firm of Aspinall, Roscoe and Lace. It was at this time that he met Jane, daughter of Mr Griffies who later was to become Mrs William Roscoe. After a 7 year long acquaintanceship and a 4 year engagement, William and Jane were married on February 22nd, 1781 when William was 27 and Jane 24 years old. They were to have ten children, seven boys and three girls, of whom eight survived. William Roscoe had a wonderful rapport with his family, and it is said that the reason he never travelled abroad was that he could not bear to be away from his family for long. He must have had many opportunities to travel with so many correspondents all over the world.

In 1784, Roscoe and a Mr Daulby revived a "Society for promoting Painting and Design" and had exhibitions under the society's patronage; Roscoe submitted one or two of his own drawings. He also designed the admission ticket. He was at this stage interested in Sir Joshua Reynolds's work, and managed to get Reynolds to give lectures on the History of Art, Prints and Engraving etc. In 1785, the painter Henry Fuseli, who was also to become a lifelong friend, was invited to Liverpool.

In 1790 the Roscoe family moved from Liverpool to a country retreat in Toxteth Park. By now William was an established and successful lawyer. However, they only remained at Toxteth for three years before they returned to the city and Folly Lane where they stayed for another six years.

Roscoe's interest in Italian history prompted him to undertake original research in order to write about the life of Lorenzo de Medici. This work was published in 1796 and was later translated into Italian, French and German. Later, in 1805, he followed this work with one in four volumes about Medici's son, who became Pope Leo X.

In 1796 Roscoe gave up his position with Aspinall, Roscoe and Lace, the firm of solicitors, supposedly to retire and have more time for his other interests. However, in order to help out a friend who was in difficulties, he was sidelined into going into banking. This was another success story for him, and the firm of bankers Leyland, Clarke and Roscoe in Castle St., prospered for twenty years until the firm was forced into liquidation through no fault of the managers, who by then were reduced to Clarke and Roscoe.

In 1797 Roscoe helped to form the Athenaeum Club before he got involved with the Botanic Garden project which was to rule his life from then on.

In the meantime, in 1799, Roscoe bought a large residence called Allerton Hall which was to be his favourite home while he was affluent. He spent time cultivating the garden and gave more time to studying botany. He had at that time a private collection of botanical books second to none and later took herbarium specimens of plants growing in his garden for the Botanic Garden, in the formation of which he was to have so much influence.

The same year as his purchase of Allerton Hall, Roscoe got together with some like-minded friends with the aim of starting a Botanic Garden in Liverpool. The port, which was being developed for the expanding cotton trade, could also be used for the transport of plants sent back from voyages being undertaken at the time to many exotic places all over the world where exciting discoveries were being made. Roscoe had been studying the writings of the botanical explorers, and realised the potential of growing economically useful plants to the advantage of British commerce. Three friends, Roscoe, Dr John Bostock and Dr John Rutter, started a subscription and shares system to provide funds for their ambitious project. 330 gentlemen subscribed to the private society.

In May 1802, the Botanic Garden was established on a 10 acre site overlooking the river, next to the present University campus. A conservatory, 73 m long, one of the largest known, had five sections with different temperatures. There was a rock garden, made from the stones from ships' ballast. The Botanic Garden was opened in 1803.

Seeds, roots and plants were obtained through ships trading to America, Asia and Africa, from other botanic gardens, nurseries, private collectors and from plant collectors in the field. John Shepherd from Manchester was made Curator — a wise choice. Roscoe and Shepherd, who got on very well together, knew William Carey, a missionary in India who sent plants and seeds to Liverpool. Dr Roxburgh and Dr Nathaniel Wallich sent material from India and the Himalayas, and before long many plant exchanges were taking place. Liverpool sent fruit trees to Calcutta. In return for plants sent to St. Petersburg, Liverpool received Pallas's *Flora Rossica*. The Liverpool Botanic Garden was run by the Liverpool Botanic Committee with Roscoe as President; the members were citizens whose aim was strictly botanical. By 1808, thousands of different types of trees and shrubs were established in the open, with many further species under glass, and students came to study the plants from all over the country.

George Bentham mentions meeting the Curator, John Shepherd, at the Botanic Garden in Liverpool and had a high opinion of him and the gingers which were kept in the conservatory….. "of which no such collection exists anywhere; I mean the Scitamineae of Jussieu, Monandria of Linnaeus. Some of the Hedychiums in flower were the grandest things I had ever seen, the Alpinias, Zingibers, Globbas, Roscoeas etc. were all remarkably fine, particularly *Alpinia nutans*".

While working in banking Roscoe had revived his love of botany, kindled when he was young, and had taken up the subject again seriously; he was able to study living plants at the newly established botanic garden. He had come to the attention of Sir James Edward Smith, who in recognition of Roscoe's interest in the gingers (*Zingiberaceae*) decided to dedicate the first volume of his *Exotic Botany*, published in 1804, to him. Later, Smith was to give the name *Roscoea* to a new genus of gingers found in Nepal by Dr Francis Buchanan, after his friend William. The genus was featured in the second volume of Smith's *Exotic Botany* with a painting by James Sowerby, executed in 1806, of the type species of the genus, *R. purpurea*. William was very pleased to hear of the new genus, but hoped that "this nymph of the Asiatic mountains will, like a faithful spouse, retain the name which you have imposed on her and not, like many of her sisterhood, elope to some more favoured admirer". Roscoe need not have worried; 200 years later the name has not been changed.

J.E. Smith was, in 1788, the founder and first President of the Linnean Society, and was later to purchase the entire collections of the Swedish botanist, Carl Linnaeus, who introduced the binomial system of classification. The collections were later to be housed in the Society's premises off Piccadilly in London.

William Roscoe Esq., R.S.L. – F.L.S. Painting by J. Lonsdale Esq. Engraved by S. Freeman. Fisher, Son & Co. London, 1847.

Roscoe was a follower of Linneaus and on 1st November 1803 was nominated as a Fellow of the Society, and was elected on 17th January 1804. He journeyed to London for the occasion and was able to meet Sir Joseph Banks. Around this time, Roscoe's interest in agriculture saw him involved with the reclamation of Chat Moss, an area of land between Liverpool and Manchester. Many years later it was a farm at Chat Moss to which Roscoe escaped in order to avoid arrest when the banking business failed.

During his banking days he was persuaded to stand for Parliament, and he became an MP for Liverpool in 1806, although he stepped down in 1807. In 1787, Roscoe had written a long poem called "The Wrongs of Africa" and it was his views on, and hopes for, the abolition of the slave trade that he wished to be able to put before Parliament. This was a tricky subject in Liverpool, where the prosperity of the city was based on the slave trade and other forms of trade at that time. Roscoe worked tirelessly for the abolition of the slave trade between 1788 and 1824. The abolition vote had come while he was an MP, but it was not until after Roscoe's death that the practice of slavery was finally abolished in 1833. He also worked towards the Reform Bill which was not to materialise until 1832, after Roscoe's death. While an MP, Roscoe was obviously popular; slogans appeared such as: "Roscoe and Integrity", "Roscoe for ever" and "The friend of the People". In those days his political persuasion would have been described as Whig. Roscoe was a patriot but was not afraid to criticise things going on in his country of which he disapproved.

In 1806 Sir James Smith encouraged Roscoe to present a paper to the Linnean Society. It was entitled "A New Arrangement of the Plants of the Monandrian Class usually called Scitamineae". It seems to have been read over two days, April 15th and May 6th. The paper was then published in the Transactions of the Linnean Society, volume VIII, pages 330–336, in 1807. By then Roscoe declared that there were 53 species, 23 of which he had been able to study from live material. Encouraged by the success of this presentation, Roscoe continued to publish papers on various botanical subjects but his finest was not to be completed until the very end of his life.

Johann Reinhold Forster's herbarium was purchased from the University of Halle; these plants were collected during Captain Cook's second voyage around the world (1772–1775). From 1805 Roscoe was also involved with expanding the herbarium at Liverpool, and drew on his many botanical friends to help in this. In 1806 he obtained for Liverpool the herbarium of Thomas Velley, a botanical friend of Smith's. Duplicate specimens from Smith also reached Liverpool.

Roscoe was always writing poetry, and around 1807/1808 he composed a poem especially for his son Robert called "The Butterfly's Ball and the Grasshopper's Feast". This seemed to have universal appeal and came to the attention of George III who asked to have it set to music for his own children.

Between 1814 and 1821 Roscoe became involved with the arrangements for opening the Royal Institution on 25th November 1817. He first became its Chairman, and was made President in 1822. It was set up as an adult education establishment. He had already been made a Freeman of Liverpool in 1815, even though his views were not always in line with the general consensus. His way of winning over people to his way of thinking, which he considered to be moral, was by persuasion and effort.

The failure of the banking business on 6 February 1816 forced the sale in that November of many of his rare books, prints, manuscripts, objets d'art and paintings. These he had been collecting fervently during the years of his marriage, even though at times the family could not afford his indulgences. Their sale must have been a severe blow as by this time William was a well known authority on Burns and Pope, and famous abroad as well as in his own country. However, even though his business activities ceased, Roscoe still remained a powerful authority. Many of his paintings found their way, through the generosity of his friends, to the Walker Art Gallery where they can be seen today as the "Roscoe Collection". Many very important early Italian and Flemish masters are saved for the nation through Roscoe's astute artistic flair and knowledge in the original purchase of these works of art. One major purchaser of paintings from Roscoe's collection was Mr T.W. Coke of Holkham, who was to become 1st Earl of Leicester in 1837.

It was during this rather difficult time for Roscoe that he corresponded on botanical matters with Sir William Hooker (later Director of Kew), who at that time was Professor of Botany at Glasgow. The letters are housed in the Liverpool Reference Library. Roscoe also had another important paper heard by the Linnean Society called "Remarks on Dr Roxburgh's description of the Monandrous Plants of India". This was published in 1816 in the Transactions of the Linnean Society London volume XI, pages 270–282.

In 1820 poor Roscoe was declared bankrupt when creditors pressed for payment. His bank had got into difficulties as a result of the economic changes that followed the ending of the war with France. Again his influential friends rallied round and raised funds for him, improving his position. The family moved to a house in Lodge Lane, which was to be his last home. His wife, Jane, died in 1824.

Even in old age and in poorer circumstances, Roscoe wanted to help those around him. In 1826, Roscoe and his circle of friends assisted the great American ornithological artist, John James

Audubon, to exhibit at the Royal Institution. The artist sent Roscoe a drawing when he left Liverpool with the note "for ever yours, most devotedly attached friend". During the last years of his life Roscoe took an interest in Prison Reform and wrote at length on the subject. He also at this time formed the "Liverpool Association for superceding the Use of Children in Sweeping Chimneys" — yet another worthy cause. It was said of him at the time "Whatever Roscoe did, he did it with distinction". He made the Liverpool of his day famous for many cultural pursuits and he has left a legacy of volumes of poems written by him and other members of his family, giving an insight into the political and cultural situations of that era. It has been said that Roscoe was Liverpool's greatest citizen.

Roscoe and J.E. Smith corresponded until 1828 when Smith died. (The majority of these letters are in the 'Smith Correspondence' at the Linnean Society of London; the rest are in the Liverpool Reference Library). It must have been sad for Roscoe; not only had he lost a wonderful friend, but Smith had been encouraging Roscoe for many years in connection with his *magnum opus*, which Smith was unlikely to have seen in its final stages. It was after Smith died that this, Roscoe's finest work "Monandrian Plants of the Order Scitamineae" was finally published, copies of which today fetch a high price as a collector's piece. It was originally published between 1824 and 1829 in 15 parts priced at 1 guinea each; only 150 bound copies of the set were printed. At the time, Roscoe declared that the number of Scitamineous plants published in 1807 had been 50 and that the number was "now not less than 200". The work is illustrated by some 112 plates, executed by many different artists between 1824 and 1828 as the plants came into flower in the Botanic Garden. Each plate is accompanied by botanical descriptions, synonymy and appropriate text. Many letters exist (in the Liverpool Reference Library) between Roscoe and eminent botanists of the time such as Dr Wallich, J.E. Smith and Joseph Banks who had helped him during the compilation of the information needed for his work. Roscoe was by now becoming frail, but he had accomplished the publication of the "Monandrian plants…" which had been many years in the making.

On 30th June 1831 Roscoe died at Lodge Lane, of a severe bout of influenza. He was buried at Renshaw Street Chapel on 30th June the same year. His grave was discovered there during construction work in 2006; sadly the memorial stone was left alone under the new building (comm. John Edmondson, Head of Science, World Museum Liverpool). Just off Mount Pleasant is a street called Roscoe Street and close by are the small Roscoe Gardens where there is a simple sandstone memorial plate. Sadly, when I visited the garden in August 1999 the area looked very run down. Roscoe should never be forgotten for the great gestures made by him for his beloved city. An exhibition was mounted in the city to commemorate the centenary of his death.

In 1836 the Liverpool Botanic Garden moved to Edge Lane, two miles away from the original site. John Shepherd died the same year and his nephew Henry took over the curatorship. In 1841 the Liverpool Corporation took over the management of the Garden. The decline of the garden accelerated during the 2nd World War when all the glasshouses were wrecked.

I stood in St. George's Hall, one of the finest neo-classical buildings in the world, which together with the Walker Art Gallery, the Liverpool Museum and the William Brown Library form the old centre of culture in the city. One of the twelve statues in the Great Hall is of William Roscoe, in the splendid company of Sir Robert Peel and William Gladstone. The statue was sculpted in 1841 by Francis Chantrey, and was in the Royal Institution before being placed in its present position of honour in the Great Hall. Looking up at the statue I could not help but feel humbled and also proud that this very famous man of his time, with such influence, had his name firmly attached to my special group of plants.

4. MORPHOLOGY

THE GINGER FAMILY (ZINGIBERACEAE)

Roscoea is a Sino-Himalayan genus in the pantropical family *Zingiberaceae*, which comprises about 1200 species in around 40 to 50 genera. Earlier publications may treat them as part of the larger group 'Scitamineae'. The majority of the species grow as understorey, shade-loving plants, and are a major component of the tropical rain forest flora of southeast Asia. Many of the aforementioned plants are robust, some attaining six metres or even more in height. By contrast, the species of *Roscoea* grow at high altitudes in the subtropical and warm temperate zones. They are much smaller than their tropical cousins; the tallest wild *Roscoea* is recorded as attaining about one metre high, although the average height is much less. The preference of the species of *Roscoea* for high altitude sites means that they are quite at home in the north temperate climes of European, Antipodean and North American gardens.

Zingiberaceae (collectively known as 'gingers') is an economically important family. Many species contain essential oils, and their rhizomes, seeds or leaves are used as spices and aromatic herbs to enhance the taste of food. Other gingers have medicinal properties, and many others have been grown as ornamentals and for cut flowers. The following species are culinarily well known: *Zingiber officinale* (ginger), *Curcuma longa* (turmeric), *Elettaria cardamomum* (cardamom), *Kaempferia galanga* (cekur, kenkur) and *Alpinia galanga* (greater galangal, lengkuas, laos). All these, apart from the *Elettaria*, are also used medicinally, as are another species of *Zingiber*, *Z. zerumbet*, *Curcuma aeruginosa* and *Curcuma zedoaria* (zedoary).

Many of the gingers have brightly coloured bracts or flowers, although the flowers of some species can be rather short-lived. The following species are recognised as good ornamental plants for the garden or conservatory, depending on which part of the world one inhabits:

Alpinia purpurata has inconspicuous flowers but a long inflorescence of large attractive red, pink or, less commonly, white bracts.

Alpinia zerumbet (shell ginger) has a long, nodding inflorescence with large white flowers streaked with yellow and red.

Curcuma zedoaria and *C. xanthorhiza* both have an attractive terminal head of sterile, colourful, pouched bracts.

Etlingera elatior (synonyms include *Alpinia elatior*, *Elettaria speciosa*, *Nicolaia elatior*, *Nicolaia speciosa* and *Phaeomeria magnifica*, and the common names, torch ginger, porcelain rose, and rose de porcelaine) has a tall inflorescence; the flowering part, a compact cone-shaped structure with deep pink, red or purplish bracts and flowers, is held at the top of a long stalk, making it a popular cut flower.

Globba winitii has a long, nodding inflorescence with large violet bracts.

Many species of *Hedychium* are valued for their large flowers, and many of them also have a heady perfume. The best-known of these is *Hedychium coronarium* (ginger lily) with white, strongly scented flowers; its near relative *H. flavescens* (yellow ginger) has deep yellow flowers. Another species found in European conservatories is *H. gardnerianum* which has smaller yellow flowers, long-exserted red stamens and a good perfume.

Kaempferia rotunda (resurrection lily) and *K. pulchra* are small plants and are happy in the shade of larger plants; they are attractive in having beautifully marked leaves.

Riedelia corallina comes from New Guinea and has rounded, basal, short-stalked coral-red inflorescences.

Zingiber spectabile also has basal inflorescences held on long stalks; the attractive parts are the pouched yellow or red bracts. It has been used as a long-lasting cut flower.

The ginger family, *Zingiberaceae*, is divided into four tribes: *Alpineae, Zingibereae, Globbeae* and *Hedychieae*. *Roscoea* belongs to the *Hedychieae*. Its closest relatives within that tribe would appear to be the genera *Curcuma, Hedychium,* and *Kaempferia*. However, *Cautleya*, another Sino-Himalayan genus is its closest relative and shares many characters; the two genera differ in just a few characters of the leaf bases, ligules, bracts, capsules and seeds. Four species have been described in *Cautleya*; two, *C. gracilis* (Sm.) Dandy and *C. spicata* (Sm.) Baker are recognised in the *Flora of Bhutan*; the other two (*C. cathcartii* Baker and *C. robusta* Baker) are probably synonyms of the first two. The genus is in need of further detailed study.

The Tribe *Hedychieae* is distinguished by having the plane of distichy of the leaves parallel to the rhizome, asymmetric leaf bases, petaloid staminodes which are usually free from the labellum, and the ovary either trilocular with axile placentation or unilocular with basal or free columnar placentation.

DESCRIPTION OF THE GENUS ROSCOEA

Perennial herbs up to 1 m high, the erect, annual leaf shoots arising from a small, sympodial, underground rhizome which serves as a resting organ, and which bears spindle-shaped or ellipsoid swollen roots (see Plate 19 illustrating *Roscoea praecox*, and Fig. 1). Similar tuberous roots are also found in *Globba, Curcuma* and some *Kaempferia* species, and in all these the rhizome segments are greatly reduced. In other genera, which do not have swollen roots, the rhizome is generally much longer and/or thicker. It would seem that the tuberous roots function as storage organs, when the upper part of the plant becomes dormant during the cold and/or dry seasons that occur in the regions where these genera grow.

The leafy stem is formed of closed, overlapping leaf sheaths (Spearing, 1977), distichously arranged or forming a rosette, but never spirally arranged. In certain species such as *Roscoea purpurea* the leaf sheaths can be flushed purple or red. The linear, lanceolate or oblong-ovate broad-bladed perfect leaves, up to eleven in number, which have conspicuous midribs, are usually formed close together and are generally green but occasionally brownish. The leaf veins are parallel, or diverge from the distinctive V-shaped keel. The basal sheathing leaves, which can be up to five in number, lack blades or have only blade rudiments which become larger on the leaves further up the stem. Each new leaf grows inside the sheath of the preceding one. The asymmetric base of the lamina, which in some species may be auriculate, is not differentiated from the sheath, but the insignifant

Fig. 1. Roots of **Roscoea schneideriana**. Cultivated and photographed by John Fielding, Sheen.

ligule is positioned here on the adaxial leaf surface, passing across the base of the leaf blade or sometimes the petiole-like part of the lamina. The surface of the leaves is usually glabrous but *R. ganeshensis* has hairy-leaves, and there are also hairy varieties of some other species.

The terminal, central, erect inflorescence is strictly a monochasial cyme but is normally referred to as a raceme or spike. The peduncle is short and is enclosed by or, sometimes, exserted from, the leaves. Unlike many gingers, the species of *Roscoea* flower seasonally. The flowers appear precociously or with the leaves and are short-lived, being at their best for no more than 24 hours. They open in succession, and may be white, pink, red, purple or yellow, and sometimes a combination of two of these colours. The flowers are zygomorphic (with bilateral symmetry), and are borne singly on the main axis of the inflorescence. Each is subtended by a bract; these bracts are spirally arranged. The bracts are oblong to elliptic, membranous, and green to whitish, and are either free, or the lower 1–3 are fused along their hyaline margins, splitting apart at maturity. The membranous calyx is tubular with an oblique mouth, and is bi- or tri-dentate at the apex. The calyx is attached below the inferior ovary.

The flower is made up of a single dorsal petal and two lateral petals. The dorsal petal is erect, circular, obovate, obcordate or elliptic in outline, concave, cucullate (hooded) towards the apiculate apex. It may have a basal claw. The equal lateral petals are usually free, linear-oblong or elliptic, and narrower than the dorsal petal. There are two petaloid staminodes, one each side of the dorsal petal. The staminodes are simple except in *Roscoea schneideriana*, where there is a smaller flange attached on the longest side (see Fig. 47). It is not known if this feature is present in all plants of this species. The staminodes converge and are erect, circular to elliptic, asymmetrically obovate, spathulate or rhombic, veined; they are usually clawed to some extent. The staminodes are free from the simple,

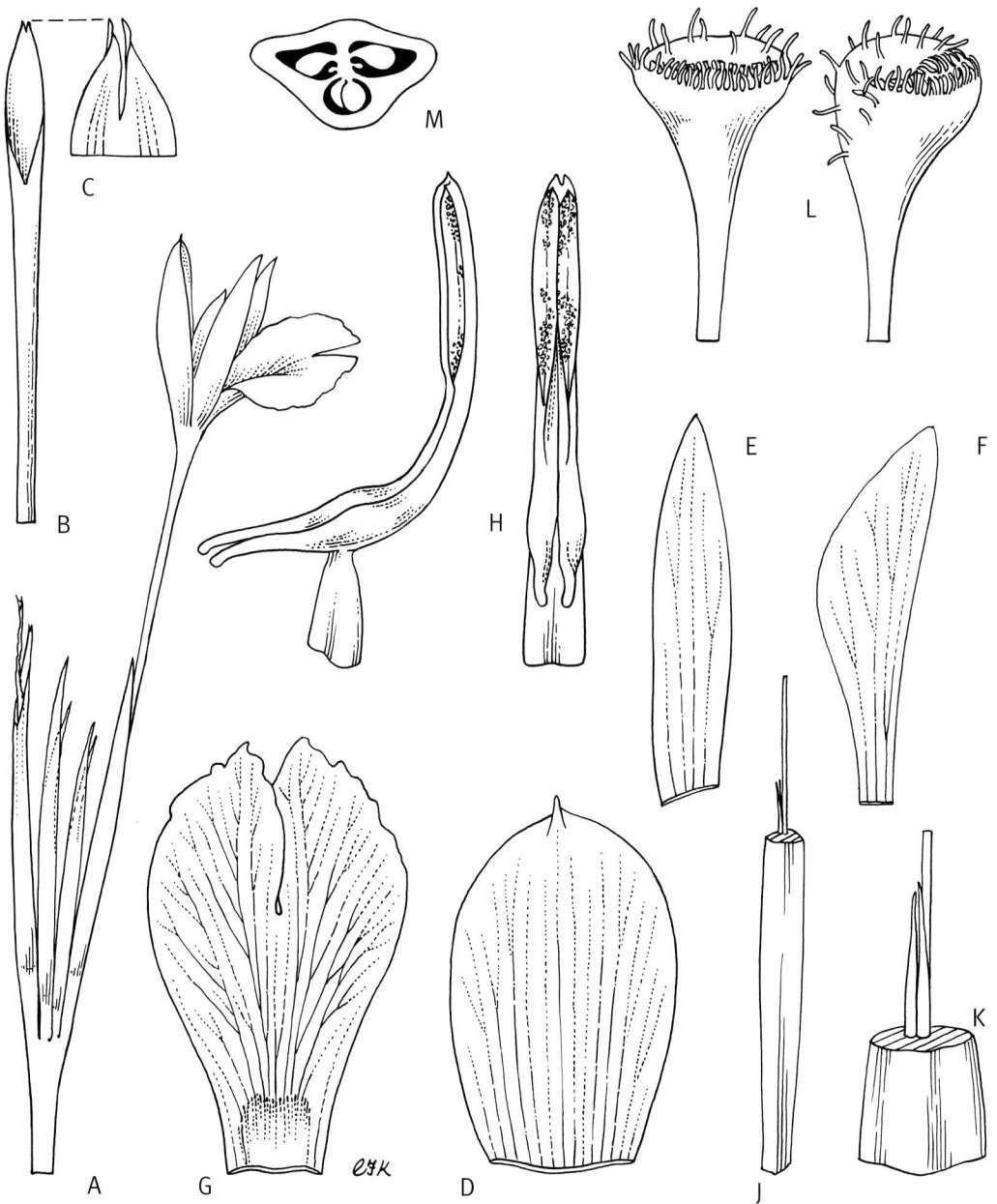

Fig. 2. Dissection diagram of **Roscoea praecox** to illustrate floral parts, drawn by Christabel King. **A** inflorescence; **B** bract; **C** detail of apex of bract; **D** dorsal petal; **E** lateral petal; **F** staminode; **G** labellum; **H** stamens, front and side views; **J** ovary, style & epigynous glands; **K** detail of ovary and epigynous glands; **L** apex of stigma; **M** transverse section of ovary.

Fig. 3 (left). ***Roscoea schneideriana*** showing closed seedpod held above leaves. Cultivated at R.B.G., Kew and photographed by Jill Cowley, R.B.G., Kew.

Fig. 4 (right). ***Roscoea cautleyoides*** showing open seedpods held above the leaves. Cultivated at R.B.G., Kew and photographed by Richard Wilford, R.B.G., Kew.

emarginate or bilobed labellum, which is usually the largest of all the petal-like parts of the flower. The labellum is usually obovate, sometimes deflexed and may be clawed or not. The origins of the labellum are believed, by some botanists, to be originally two to three modified sterile stamens but not everyone agrees on the derivation of the labellum and the staminodes. Rao *et al.* considered the labellum to be a double structure formed from the two united antero-lateral staminodes of the inner whorl. All of the perianth segments arise from the long, slender perianth tube which is attached to the apex of the ovary and is usually longer than the calyx (see Fig. 2).

There is one fertile stamen on the same radius as the dorsal petal. It consists of a short, flat, erect filament and a spurred, versatile uncrested anther which dehisces by longitudinal splits. The short, free sterile appendages, or spurs, at the base of the anther curve away from the narrow thecae and are pointed except in *R. schneideriana* where the ends are white and rounded (see Figs. 48, 49 and Plate 11). The anther thecae are usually at right angles to the spurred appendages; the basal connective elongation is either short or long. The long, filiform style is held in a furrow between the two anther thecae and the appendages, and is topped by the stigma which is exserted just above the anther. Where

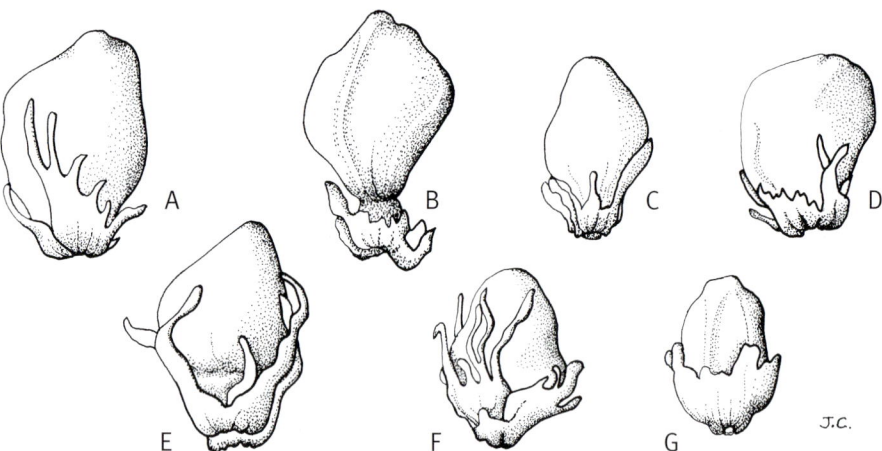

Fig. 5. Drawings of seeds by Jill Coombs. A ***Roscoea purpurea***; B ***R. auriculata***; C ***R. wardii***; D ***R. scillifolia***; E ***R. cautleyoides***; F ***R. humeana***; G ***R. forrestii***.

the base of the style is attached to the ovary, there are two thin, erect, linear, epigynous glands (stylodial nectiferous glands). There is general agreement that these epigynous glands are outgrowths of the upper surface of the ovary, and are not staminodes or stylodia. In some other genera, although not within the *Hedychieae*, the glands are hardly noticeable. The ovary is cylindrical and trilocular with axile placentation and numerous ovules. The stigma is usually erect and galeate but in *R. schneideriana*, it is uncinate and infundibuliform (Fig. 48C). In both forms there is a fringe of hairs at the aperture.

The fruit is an elongated, ellipsoid, fleshy capsule which is tardily dehiscent (Figs. 3 & 4). The valves separate at the apex and divide the capsule into three parts that fold back to reveal the seeds (Fig. 6). The seeds are spherical, ovoid, elliptic, or sometimes turbinate or angled; they are smooth or rugose and when mature may be greenish, brown or blackish. Each seed has a fleshy, lacerate, white aril (see Fig. 5). The agents responsible for pollination and seed distribution are poorly known, but it is generally believed that the seeds may be distributed by ants.

Fig. 6. ***Roscoea tibetica*** showing open seedpod with seeds at ground level. Cultivated and photographed by John Fielding, Sheen.

5. CYTOLOGY, ANATOMY, PALYNOLOGY & PHYLOGENETIC STUDIES

There have been few studies of the cytology, anatomy and palynology of *Roscoea*. By far the most comprehensive study of these aspects is in the area of phylogenetics, and has been carried out recently by the Thai botanist, Dr Chatchai Ngamriabsakul for his as yet unpublished PhD thesis entitled "The systematics of the Hedychieae (Zingiberaceae), with emphasis on *Roscoea* Sm.". In this chapter I quote extensively from Ngamriabsakul's thesis, and from his related paper "Phylogeny and Disjunction in Roscoea (Zingiberaceae)", written with M.F. Newman and Q.C.B. Cronk and published in the *Edinburgh Journal of Botany* volume 57(1): 39–61 (2000).

CYTOLOGY

Somatic chromosome counts for *Roscoea* have been published by Sharma & Bhattacharyya in 1959; Malik in 1961; Bhattacharyya and Bisson *et al.*, both in 1968; Mahanty in 1970; Mehra & Sachdeva in 1971, 1976, & 1979; Chen *et al.* in 1986 and 1987 and West & Cowley in 1993. The basic chromosome number is $2n = 2x = 24$ and the chromosome morphology is uniform, the size being 1-2 µm for all species and populations examined. One instance of polyploidy in an unidentified species has been recorded, $2n = 48$, and there is one count of $2n = 26$ for *R. purpurea* (Mahanty, 1970). In the latter, the drawing of the chromosomes indicated that two of them had probably separated into chromatids at late metaphase, giving rise to the false impression of a somatic number of $2n = 26$.

Dr Ngamriabsakul included a cytological study in his thesis (2001). He reported *Roscoea alpina* as having $2n = 26$ and *R. purpurea* and *R. auriculata* as having $2n = 24$. He collected and fixed root tips around mid-day, because both Lim in 1972 and Newman in 1990 had found that in the *Zingiberaceae*, there were likely to be large numbers of dividing cells at a stage where counting is easiest, at this time of day. The chromosomes of the species studied were mainly metacentric. *Roscoea* chromosomes are smaller than those of *Kaempferia*, but larger than those of *Curcuma*; all three genera are part of *Hedychieae*.

Ngamriabsakul suggested that the differences in his counts from previous studies of the same species could be due either to centric fission of one of the pair of chromosomes, or due to centromere breakage without reunion, giving 2 telocentrics or iso-chromosomes.

ANATOMY

P.B. Tomlinson published papers on the anatomy of the *Zingiberaceae* (1956), the phylogeny of the *Scitamineae* (1962) and the *Zingiberales* in 1969, but made no specific reference to the genus *Roscoea*. In 1991 two Chinese botanists, Guo Yi and Wu Qigen, published a paper, in Chinese, on the anatomy of the genus, entitled "The comparative anatomy of the vegetative organs in *Roscoea* Smith (*Zingiberaceae*)".

They studied nine species, mainly Chinese, and described the structure and distinguishing features of the leaves, flowering stems, rhizomes, tuberous roots and fibrous roots. The authors divided the nine species into three morphologically defined groups, A, B and C. Each of these groups was defined by a syndrome of anatomical characters, as displayed in the tables that accompany the paper. They found that some leaf characters appeared to be related to habitat, and that various characters can be interpreted as adaptations to high altitude seasonally dry environments. Such are the closed leaf sheath, the semi-parallel straight veins, the lack of an abaxial vascular system and the presence of an adaxial leaf vascular system, more than one layer of hyperdermis, the larger stomata, the presence of coarse crystal sand, the vestigial rhizomes and the well-developed tuberous roots. They concluded that *Roscoea* is a natural taxon, and that it represents an advanced branch of *Zingiberaceae*.

PALYNOLOGY

In 1988, Liang Yuan-hui published a paper in *Acta Phytotaxonomica Sinica*, that gave an overview of studies undertaken on the pollen morphology of the family *Zingiberaceae* in China and their relevance to taxonomy.

The studies used both light microscopy and scanning electron microscopy. The pollen type, grain shape, aperture type, wall thickness and ornamentation is listed for each taxon; each is illustrated by a photograph. The author found that the pollen of *Roscoea* is of the spheroidal, nonaperturate type, in the spinate subtype, and the long-spinate group. In other words, the pollen grains are spherical, lack pores, and the wall bears long spines. Three other genera in the *Hedychieae*: *Boesenbergia*, *Caulokaempferia* and *Cautleya*, also fall into this group. Eight species of *Roscoea* were included in the study; photographs of their pollen types at varying magnifications appear in their Plate 3.

The sizes of the pollen grains are given in a table. The wall thickness varies from 1.6 to 3.5μm. The spines on the grain wall are described as spinate if they are over 3.0μm long, and spinulate, if less.

We know very little of pollination and pollinators in *Roscoea*. Indeed, little is known of pollinators within the whole family *Zingiberaceae*, and there is surely a potentially open field for researchers with access to the living plants in their natural habitats. T.B. Fletcher and S.K. Son, in a paper on veterinary entomology for India in 1931, suggested that long-tongued flies pollinated *Roscoea*. This was repeated by W. Dierl in 1968 in a paper on *Corizoneura longirostris*.

Ngamriabsakul, in his thesis, suggests that features of the floral structure of *Roscoea*, such as the basifixed versatile anther and the appendages, indicate that the pollinators are bee species that are attracted to the flower for the nectar. "The pendulous lip of the flower is thought to be a platform for the pollinator to enter and in doing so the appendages will be pushed, bringing down the anther into contact with the back of the pollinator. Fruits of *Roscoea* are often observed in the Royal Botanic Garden, Edinburgh where there is probably no true pollinator of *Roscoea* as in its wild habitat. Garden bees may be pollinating the flowers, leading to the formation of fruits. Because *Roscoea* grows in clumps of individuals, possibly other insects or wind may also play a part in the pollination." In 1881, a Mr R. Irwin Lynch, Curator of the Botanic Garden in Cambridge at the time, described cross fertilisation in *Roscoea purpurea* to the Linnean Society, comparing the general form of the flower to that of *Salvia grahami* which has a similar structure. He observed that not only the anther but also the stigma moved down towards the insect as a result of the insect's weight acting on the labellum. He was struck by the great similarity in the pollination mechanisms of two plants from completely different families.

There is also very little information on seed dispersal in the genus. It would seem logical that the lacerate arils of the seeds make them both attractive to ants and also easy to grasp and carry away. This theory is expanded upon in a paper written by Charles Robertson in 1897 in the *Botanical Gazette*, but there is no specific mention of *Roscoea*.

PHYLOGENY

Cladistics is the term given to a method of classification that relies solely on monophyly for classification (i.e., all species in each group are descendants of a single common ancestor). Given a set of organisms and a set of characters in use for classification, this objective method will ideally give the same results as systematics. The characters that are used for grouping are shared derived ones. The principal concept is the parsimony of evolution or the requirement of minimum changes in the course of evolution. It means that the shortest hypothetical pathway of change that explains the present pattern of data used in the systematic study is considered to be the most likely evolutionary route.

In plants there are three different kinds of DNA in a cell: nuclear DNA, chloroplast DNA and mitochondrial DNA. A phylogenetic study of *Roscoea* was undertaken using sequence data from the internal transcribed spacers (ITS) of the nuclear ribosomal DNA (nrDNA). The ITS sequences are well established as being useful in systematics. They have rates of substitution (i.e., number of changes in DNA per unit time) that are useful for evaluating generic and species level relationships in plants. ITS regions have proved to be useful for studying the evolutionary relationships of the family at the species level.

Ngamriabsakul aimed to test whether *Roscoea* is monophyletic (i.e., are all the species derived from a single common ancestor). He also aimed to test whether *Roscoea* and *Cautleya* really are closely related, as the similarities in their morphology had suggested. The study used the living collections in the Royal Botanic Gardens at Edinburgh and Kew. DNA was extracted from a total of 16 species. It was hoped that by combining phylogenetic information from the ITS regions with distribution records and information on geological history, some insight might be gained into the evolution of *Roscoea* and its sister genera and into how a tropical family has colonised temperate regions. Alongside the analysis based on ITS sequences, he also undertook various analyses using both ITS sequences and morphological characters. Seventeen morphological characters, for all 19 species known at the time in the genus, were scored.

Two species of *Cautleya*, *C. gracilis* (Sm.) Dandy and *C. spicata* (Sm.) Baker, and two species of *Curcuma*, *C. amada* Roxb. and *C. parviflora* Wall. were used as outgroups. The results from both the ITS-based and morphology-based analyses suggest that *Roscoea* is monophyletic with the genus *Cautleya* as sister group (i.e., the closest related genus). The DNAs of *Cautleya* and *Roscoea* are not very different, and this suggests a close relationship between the two genera. This is supported by their similar morphology and overlapping distribution area. *Cautleya* is not only found with *Roscoea* at lower levels of the Himalaya and in south central China but, like *Roscoea*, is also recorded from high altitude sites on tropical mountains in Burma, and in the north of Thailand. However, the present distribution of *Roscoea* and *Cautleya* is centred on Assam.

Roscoea itself is divided into two sister clades (i.e., groups, each derived from a single common ancestor) which correlate with geography: a 'Chinese' clade, and a 'Himalayan' clade. These two groups are mutually exclusive, and are separated by the 'Brahmaputra gap', a region in which no *Roscoea* species have been recorded. The Chinese clade comprises seven species from China and one from Burma. The Himalayan clade includes seven species from the Himalaya.

It is possible that *Roscoea* originated in Assam and spread east and west along the nearest mountain ranges, thus accounting for the separate Chinese and Himalayan groups. This is supported by a single maximally likely tree showing that a clade of *Roscoea* plus *Cautleya* shares an ancestor with the *Hedychium* species clade. However, the phylogenetic analyses did not suggest that either the Chinese or the Himalayan clade could be considered most closely related to *Cautleya*; they are sister groups.

All of the species in the Chinese clade (except *Roscoea australis* from Burma) are found in Yunnan province (mostly in Lijiang and Dali) and some extend to parts of Sichuan, according to Ngamriabsakul (2001). The data suggest that this is an area of rapid evolution of a complex of *Roscoea* species. On the other hand, the area of greatest diversity of the Himalayan clade is in central Nepal. In particular, Ganesh Himal is an area in which grow five species out of the eight in the entire Himalayan region.

The distribution of *Roscoea* is strikingly discontinuous. There are no records from that part of Assam where the Brahmaputra river flows south around the eastern end of the Himalayan chain. This gap in the distribution coincides with the boundary between the Chinese and the Himalayan clades. Although it is possible that the Brahmaputra gap is an artefact, caused by undercollection, it may also represent a real phytogeographical boundary. The region of the Brahmaputra gap is known to be undercollected, because the area has historically been difficult to access. Northeast India has only been surveyed casually, and more observations from this area are badly needed. Of course, the area may really have no species of *Roscoea*. Although the Himalayan mountains form a continuous, geologically connected chain, here the eastern Himalaya rise rather abruptly from the plain and the sub-Himalayan zone is very narrow or absent. This abrupt rise of the mountain range and its horseshoe shape may serve as a barrier between the two sides of the area. Thus the disjunct distribution of *Roscoea*, between two sides of north-eastern India, may be genuine along with other examples of Indian disjunctions.

6. ECOLOGY AND CONSERVATION

The genus *Roscoea* occurs mainly in the Himalayan range, extending into Tibet (now known as Xizang) and into western China in Yunnan and Sichuan Provinces, and with outlying stations in the Khasi Hills of Meghalaya, India, and the Chin Hills of Burma. (In this book I use the term Tibet in its old sense, referring to the areas bordering the northern Himalaya as far west as Kashmir, and eastwards through Nepal, Sikkim, Bhutan, north Burma, Assam, Yunnan and Sichuan). These are all regions with a monsoon climate, although this is usually modified in mountain areas. Species of *Roscoea* have been found at altitudes between 1000 and 5000 metres.

The Tibetan plateau was beneath the sea in the Cretaceous period and was uplifted after the Tertiary as the continent of India drifted northwards and collided with the Asian land mass. It was elevated in a series of east to west folds. The great rivers draining the eastern plateau — the Tsangpo (which becomes the Brahmaputra), the Irrawaddy, the Salween, the Mekong, and the Yangtze (Jinsha Jiang) — in the Chinese provinces of Yunnan and Sichuan, cut deep parallel gorges through the mountain chain as the land was uplifted.

During the Quaternary Ice Ages, the Tibetan Plateau was under intense glaciation. The flora was driven southwards by the advance of ice, but the existence of the great river gorges meant that it could also move up and down the mountains. Within the Plateau area, the rivers run mainly east-west and do not therefore act as a barrier to the east-west spread of the flora. However, further east, in the northeast corner of India, the rivers break through the chain from north to south and thus act as a barrier to the east-west spread of plants. As a result of the uplift of the Himalayan ranges and their glaciation, the flora of the eastern Himalaya has affinities with western China across the gorge country, the Sino-Himalayan flora having spread east to west from the river gorge country and penetrated far into Tibet. Floristically the Sino-Himalayan region comprises the whole of the Tibetan plateau, the great Himalayan range, the river gorge country and Chinese Tibet.

The climate of the region is monsoonal, with a long summer wet season and drier winters. Much of the rain falls between May and October; the total amount varies enormously from place to place, with annual totals of as much as 4000 mm in the Khasi Hills of Meghalaya (*Roscoea brandisii*) but much less at Lhasa on the Tibetan Plateau (*R. bhutanica*). To the west, in Kashmir, the mountain chain swings further to the north, and the rainfall tends to fall more in the winter; here *Roscoea* is scarce or absent.

In the Sino-Himalayan region, the vegetation tends to form distinct altitudinal zones. The lowest slopes of the mountains bear (or used to bear) tropical evergreen forest; with increasing altitude the composition of the forest changes and species belonging to temperate families and genera become more prominent. Broad-leaved trees tend to be replaced by conifers, and eventually an altitude is reached at which trees can no longer survive. In this 'alpine zone' the vegetation is herbaceous. Of course none of these dividing lines is sharp and the succeeding zones grade into one another in mosaics and patchworks of different vegetation. It is this environment which *Roscoea* prefers, around the lower reaches of the alpine zone. The forested areas of the lower gorge country are favoured; here the forests are predominantly of conifers, oaks and birches.

WESTERN HIMALAYA

The regions occupied by *Roscoea* can be divided into three. The first is the main Himalayan Range from Kashmir in the west to the Brahmaputra in the east. *Roscoea alpina* Royle and *R. purpurea* Sm. grow in the western Himalaya where the winters are colder and wetter than further east in Nepal; these species also have to withstand hotter and drier summers where there is only a light summer rainfall. These two species have a wide distribution extending all along the Himalayan range (Maps 1 & 2). They would appear to be tolerant of a wide range of environmental conditions. However, the majority of the species in the 'Himalayan group' come from Nepal where most of the rain falls in summer. Unlike the western Himalaya, Kathmandu, the capital of Nepal has dry winters, without snow. Here, where many *Roscoea* species are found, the monsoon rains fall between May-June and the end of September, so that the summers are wet (Fig. 7). Areas of Nepal with low rainfall have a flora which resembles that of the Tibetan plateau; in such areas, north-facing slopes have a richer flora than south-facing ones, which get more sun.

Eastwards in the Himalayan Chain, the genus *Rhododendron* becomes more prominent; the further east one travels, the more *Rhododendron* species there are. This east Himalayan flora and vegetation is very similar to that of southeast Tibet and western China. The Tsangpo (Brahmaputra)-Salween river, which breaks through the Himalayan Chain in a deep gorge, appears to divide *Roscoea* into two groups, the Himalayan group to the west of the gap, and the Chinese group to the east (see Maps 2 and 3). While this gap, which lies in the Indian Province of Arunachal Pradesh and is hereinafter referred to as the Brahmaputra Gap, in all probability represents a real distributional gap, it has to be remembered that it is also an area which is very under-collected. The reality of these two groups is backed up by the evolutionary evidence in DNA studies carried out recently (2001) by Dr Chatchai Ngamriabsakul.

Fig. 7. Cloud forest at 3250 m, Ganesh Himal, Nepal. Photographed by Bill Baker in 1992.

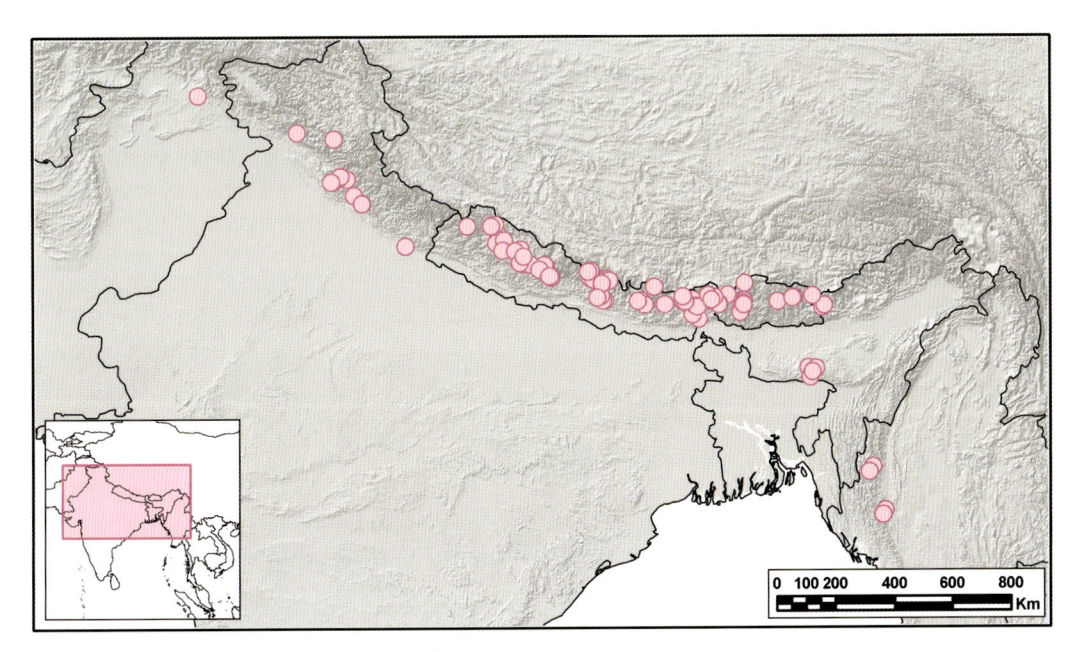

Map 2. Distribution of the Himalayan group of species.

THE TIBETAN MARCHES — SICHUAN AND YUNNAN

To the east of the Brahmaputra Gap is the area of the Tibetan Marches, lying between the great north-south rivers of China in Sichuan and Yunnan Provinces. The substrates found in the areas around the great rivers include siliceous limestones, red sandstones, conglomerates, black slates and brown shales; all these rocks have been affected by the major earth movements that produced the uplifted Himalayan Chain. To the east of the Yangtze river there is a high limestone plateau dissected by north and south flowing rivers.

Glaciation of this region left wide valleys and lakes which support a rich herbaceous alpine flora; China's temperate flora is the richest in the world. This area falls within the Eastern Asiatic floristic region, which is divided into two floristic provinces. *Roscoea* occurs in the Sino-Himalayan Province. Here rhododendrons, *Meconopsis* and *Nomocharis* are among the treasures of the flora; there are relatively few woody plants because of seasonal droughts. The winters are dry, and the plants flower in the early summer and grow during the summer rains. They then become dormant and may lie under snow for long spells in the winter.

Here *Roscoea* species are found mainly in the magnesium limestone areas, on west facing slopes, at altitudes between 1520 and 4270 m. They are associated with conifer forests dominated by *Pinus yunnanensis* Franch. and to a lesser extent *P. armandii* Franch., which are found between 2400 and 3050 m. The herbaceous flora beneath the pines includes genera such as *Cypripedium*, *Arisaema*, *Hemerocallis*, *Pleione*, *Morina*, *Stellera*, *Lilium* and *Roscoea;* the main flowering season is in May and June. The melting snows provide moisture on the steep slopes, and in open situations one can find oaks and rhododendrons. Further down, where there is more surface water, *Primula*, *Gentiana*, and *Paeonia* grow, while in the meadows genera such as *Incarvillea*, *Iris*, *Meconopsis* and *Pedicularis* thrive.

THE GENUS ROSCOEA
ECOLOGY AND CONSERVATION

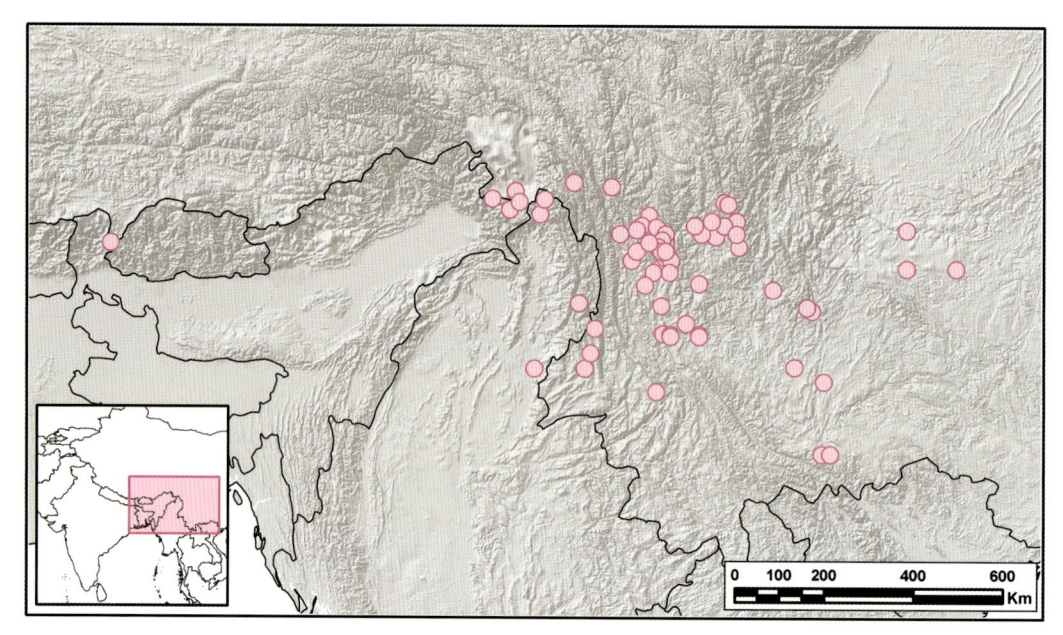

Map 3. Distribution of the Chinese group of species.

In Muli the seasons are a little different to the area around Lijiang in Yunnan; there is a cold, dry season in May and June before the rainy season between July and August, followed by a warm, dry season during September and October. The area around the limestone Jade Dragon Mountains or Yulong Shan, where Lijiang is situated, is a haven for plant hunters (Fig. 8). The highest peak of the range is 5500 m high, and the lower valleys are rich in plants. The Gang ho ba is a dried out river bed where the plants growing in the valley thrive on the white substrate (Fig. 9). One would not normally expect to see acid-loving plants growing in these limestone areas, but the heavy monsoon rains have in the long term leached the surface soil to produce acid pockets in which the rhododendrons grow.

Species of *Roscoea* have been found in several localities in the Yangtze River area of Yunnan. They are found to the north, on the Zhongdian plateau, home of the Yi people, at the edge of the Tibetan plateau, at altitudes of 3260 m and above (Fig. 10). Several species can be found in the Lijiang area mentioned above, which includes the Gang ho ba, the Lijiang Plain (Fig. 11) and Bai Shui river valley, at 2500 to 3300 m, this area being the home of the Naxi people. There are also several species of *Roscoea* in the Dali area of the Cangshan mountain range south of Lijiang, which has altitudes up to 4250 m, and which is inhabited by the Bai people (Fig. 12).

THE NORTH BURMA TRIANGLE

Kingdon Ward considers north Burma to be a western extension of Yunnan, it having a similar flora northwards, the mountain ranges merging into the Tibet plateau above the Irrawaddy basin. He called the area between the eastern and western branches of the Irrawaddy the North Burma Triangle and considered the area to be isolated botanically. North Burma and northeast Assam make up part of the Sino-Himalayan plateau which stretches for 2000 miles across Asia. The plateau has been deeply and widely eroded by glaciation, geologically recently, in the Pleistocene and this initiated changes in the climate producing the vegetation types which exist now. It has as a result produced a rich and varied flora. *Roscoea* is confined to the phytogeographical region of the alpine Sino-Himalayan zone. The Adung valley and other areas where Assam, Burma and Tibet meet in the upper Irrawaddy and where Kingdon Ward collected *Roscoea wardii* shows Himalayan and Tibetan influences. As far as *Roscoea* is concerned the species found in this region belongs to the Chinese group as mentioned above. The upper Irrawaddy region is made up of conifer and *Rhododendron* forest with meadows at 2800–3050 m. There are broad open ridges where the flora has to battle with bamboo. Where the roscoea species are found the accompanying meadow flora is rich with *Primula, Lilium, Meconopsis, Androsace, Pedicularis, Corydalis, Allium, Aconitum* and *Polygonum*, reminiscent of the western chinese flora in Yunnan and Sichuan. The flora has spread eastwards from the headwaters of the western Irrawaddy into northwest Yunnan and then southwards along the high mountain ranges to find their resting place in the limestone areas. These areas are sought after by plantsmen because of the rich diversity of the flora and have been discussed above; they are home to numerous species of *Roscoea*.

OUTLYING SPECIES — KHASI AND CHIN HILLS

Roscoea brandisii grows in the Khasi (Khasia) Hills of Meghalaya (formerly part of Assam), India. These hills lie to the south of the Brahmaputra River, well separated from the main Himalayan Chain. The rainfall here is extremely high. The Khasi Hills extend eastwards and almost join up with the hills of Nagaland and Manipur. Kingdon Ward wondered why he never found *Roscoea* in Manipur, where the flora is very similar to that of the Khasi and Naga Hills of Assam and the Chin Hills of Burma.

Fig. 8. Yulong Shan mountain range, Yunnan province, China. Photographed by Jill Cowley, R.B.G., Kew.

Roscoea australis grows in an isolated area at latitude 21° N, much further south than any other locality for the genus. Mount Victoria in the Chin Hills of Burma is its type locality, but there is also a collection from Haka, some 250 km to the north in the same range. Mount Victoria is over 3000 m high and is the highest point in the southern Chin Hills which are an extension of the Sino-Himalayan plateau. There is no sign of past glaciations on Mount Victoria; the ice probably never reached so far south. *Roscoea australis* grows in the temperate semi-evergreen forest zone between 2000 and 3000 m, on sunny turf slopes where many plants belonging to temperate genera grow. As well as *Roscoea*, there are herbaceous species of *Iris*, *Anemone*, *Campanula*, *Potentilla*, *Saxifraga*, and *Swertia* as well as trees and shrubs such as *Prunus*, *Schizandra*, *Clematis*, *Malus*, *Jasminum* and *Ribes*. The flora of Mount Victoria can therefore be regarded as an outlier of that of the Sino-Himalayan region. The surrounding lowlands must have been occupied by a tropical forest flora, and the alpine flora must have reached the Chin Mountains by migrating south along the Naga-Manipur Range from the Tibetan plateau. At the time of Kingdon Ward's visit, in 1956, the hills were cultivated up to 2000 m with rice and maize, but the forest on the north and eastern slopes was intact and extended right up to the summit of Mount Victoria.

SUBSTRATE PREFERENCES

The underlying geology of the area within which the genus occurs is extremely varied. There are sedimentary rocks including limestones, sandstones, shales and conglomerates. There are also igneous rocks such as granites, and metamorphic rocks such as gneiss and schists. However, over much of the area the rainfall is high, and weathering is likely to have produced soils that are broadly similar whatever the substrate. It is clear, however, that all species normally grow in well-drained sites.

CONSERVATION

Until the ecology of plants in an area is understood, it is hard to implement conservation measures. In addition, there is a need to understand the needs of the local population and the pressures on the flora that these may generate. For instance, the logging of forests may be an important source of revenue for local people, quite apart from the role of the forests in providing fuel and building materials. For instance, I could see clear landscape changes between my first visit in 1990 on the (Zhongdian) Chungtien-Lijiang-Dali Expedition, and on my return to the same areas just two years later. On the first trip I collected three species of *Roscoea* growing together just off the Gang ho ba, a high, dry river valley near the Lijiang plain. In 1992 the area had been logged and the *Pinus yunnanensis* under which *Roscoea* had flourished no longer existed; their habitat was lost (Fig. 13). Population pressures may also encourage the clearance of forests to provide extra land for agriculture. The exercise of summer grazing rights by villagers in some areas of higher ground certainly prevents many species from flowering and seeding.

Many species of *Roscoea* are not widespread and are found only in small local populations. Near Kunming, *Roscoea praecox* is most easily seen and collected on rocky outcrops on bends in the road where the soil is red. Such habitats are by no means stable and the species growing in them may well be threatened by road widening or similar activities. Larger-scale operations such as valley

Fig. 9. The Gang ho ba, a dried up river valley, Yulong Shan, Yunnan province, China. Home to at least four *Roscoea* species. Photographed by Jill Cowley, R.B.G., Kew.

flooding for hydro electric power installations, or mining and quarrying, are a further threat to larger areas. *Roscoea tibetica* may have the best chance of survival as it can be found growing in a wide range of habitats, in open meadows as well as in pine forest, and is fairly widespread in the Burma-Yunnan borders, Yunnan itself and in Sichuan.

Earthquakes are a natural hazard to which many Roscoea habitats are prone; the whole region is in an earthquake zone. Kingdon Ward records his plight at being close to the centre of a tremendous earthquake in Assam in 1950. The place where they had camped a day or two before was completely devastated, and whole mountainsides came crashing down, damaging and obliterating habitats. This area was a known locality for *Roscoea wardii*, a species known only from a very restricted area. By chance, the Kingdon Ward collection that was later to be chosen as the type was made by him in July, just one month prior to the earthquake of August 15th. The species has not, as far as I am aware, been collected since, and who knows how many precious habitats were lost on that fateful day. One can only hope that wild populations of this species still exist. The material that is in cultivation today must have been collected by Kingdon Ward, as the area does not appear to have been searched by any other botanist since 1950. Unfortunately the provenance and collecting number have been lost on the plants in cultivation. I like to think that they are from the plants found in 1950, and perhaps from the collection chosen as the type.

As for conservation issues as regards the Himalayan group of species, the Indian Forest Service has long had conservation policies but it is only recently that Nepal has come to be aware of the need for conservation and reforestation.

Fig. 10. The Zhongdian Plateau, Yunnan province, China, where *Roscoea* species can be found. Photographed by Jill Cowley, R.B.G., Kew.

Fig. 11. Lijiang Plain, Yunnan province, China, home to *Roscoea humeana*. Photographed by Jill Cowley, R.B.G., Kew.

THE GENUS ROSCOEA
ECOLOGY AND CONSERVATION

Fig. 12. Cangshan mountain range where *Roscoea* species can be found. Photographed by Jill Cowley, R.B.G., Kew.

Fig. 13. Roscoeas, growing near the Gang ho ba in the autumn of 1990. Two years later their habitat had disappeared. Photographed by Jill Cowley, R.B.G., Kew.

ECOLOGY IN RELATION TO CULTIVATION

Writing in 1924, Kingdon Ward was a bit scathing about the roscoeas he found in Sichuan on the Muli range in the arid Litang river gorge. They grew on dry, sunny slopes with lilies, *Arisaema*, *Iris* and *Cyananthus*. Four species of *Roscoea* have been recorded from the Muli area: *R. humeana*, *R. schneideriana*, *R. tibetica* and *R. cautleyoides*. Rather strangely, Kingdon Ward called the latter " a pompous flower. It looks artificial as though cut out of paper or cheese …… not for the garden." He noticed " a dozen species on the limestone uplands…… They are nearly always overcrowded ……….. occasionally you see an ivory white one which is not so unpleasant." His remarks do not seem to have stopped gardeners coveting *R. cautleyoides* in their gardens to this day.

It is extraordinary that plants which favour rather specialist conditions, confining themselves in the wild to certain substrates and enjoying high rainfall, seem to abandon these confines when brought into cultivation and will settle down in soils, altitudes and other climatic conditions alien to those in their homeland. However, altitude, rainfall and aspect are very often more important to plant populations than the rock composition. Roscoeas grown in Europe have to adapt to cold wet winters and warm drier summers, rather than the cold drier winters and hot wet summers of their natural habitat.

7. KEY TO THE SPECIES OF ROSCOEA

No key to the species of *Roscoea* can be foolproof, because species of the genus are so variable, especially in cultivation. Herbarium specimens, which can be very useful for identification, are never all collected at the same stage of development. The gardener who wants to identify the plants growing in his garden may pick the moment when the first flower emerges to try to put a name to his treasure. Many of the species develop very quickly. A plant may start flowering with no leaves visible, and still be flowering when it is much taller and the full complement of leaves has been produced.

With this in mind, the most straightforward key available, which was devised by Ngamriabsakul & Newman (2000), has been amended here to include all the known *Roscoea* species and to include extra information that may be helpful.

1a.	Labellum longer than dorsal petal; anther appendages pointed or tapering towards tips; staminodes obliquely spathulate or circular to elliptic; thecae at right angles to or in line with appendages; flowers purple, red, pink or white, never yellow; Himalayas from Pakistan to Assam	2
1b.	Labellum mostly shorter than dorsal petal; anther appendages obtuse or globular, never really pointed; staminodes asymmetrically obovate, rhombic or elliptic; thecae at obtuse angles to the appendages; flowers purple, pink, yellow or white; southwest China, Burma or Tibet/Assam/Burma borders	10
2a.	Leaves usually 2–3(–6) at flowering time, forming a tuft; plant usually less than 20 cm high	3
2b.	Leaves usually more than 3 at flowering time, well spread; plant usually more than 20 cm high	5
3a.	Staminodes circular to elliptic	4
3b.	Staminodes obliquely spathulate	**8. R. bhutanica**
4a.	Leaves linear; first leaf auriculate; bracts obtuse	**1. R. alpina**
4b.	Leaves obovate; all leaves slightly petiolate; bracts acute	**5. R. nepalensis**
5a.	Leaves auriculate throughout; bracts equal to or shorter than calyx	6
5b.	Leaves generally not auriculate (rarely the lower leaves auriculate); bracts equal to or longer than calyx	7
6a.	Bracts exserted, equal to or slightly shorter than calyx; staminodes usually white	**7. R. auriculata**
6b.	Bracts hidden, much shorter than calyx; staminodes purple	**6. R. tumjensis**
7a.	First bract tubular, soon splitting or not; bracts and calyces ciliate	8
7b.	First bract not tubular; bracts and calyces glabrous	9
8a.	Inflorescence on on exserted peduncle, capitulate; thecae at right angles to appendages; lateral petal linear to oblong	**3. R. capitata**

8b.	Inflorescence hidden; thecae more or less in line with appendages; lateral petal elliptic	**4. R. ganeshensis**
9a.	Leaves lanceolate to oblong-ovate; dorsal petal narrowly elliptic, more than 3 cm long	**2. R. purpurea**
9b.	Leaves linear to narrowly lanceolate; dorsal petal elliptic to broadly elliptic, less than 3 cm long	**9. R. brandisii**
10a.	Leaf bases petiolate or slightly auriculate	11
10b.	Leaf bases decurrent	13
11a.	Leaves petiolate; bracts equalling calyces	**18. R. debilis**
11b.	Leaves auriculate; bracts shorter than calyces	12
12a.	Bracts acute; dorsal petal elliptic; lowest bract not tubular	**17. R. tibetica**
12b.	Bracts obtuse; dorsal petal obovate; lowest bract tubular	**10. R. australis**
13a.	Bracts longer than calyces at flowering time	14
13b.	Bracts shorter than or about equal to calyces at flowering time	15
14a.	Leaves crowded together in a fan shape; inflorescence not capitulate, peduncle hidden in leaf sheaths	**12. R. schneideriana**
14b.	Leaves rather evenly spaced up the stem; inflorescence capitulate, peduncle visible	**13. R. scillifolia**
15a.	Leaf blade abaxially glaucous; flowers deep purple	**11. R. wardii**
15b.	Leaf not as above; flowers purple, yellow or white	16
16a.	Bracts subtending flowers obtuse; lowest bract not tubular	17
16b.	Bracts subtending flowers acute to acuminate; lowest bract tubular	18
17a.	Dorsal petal obovate to obcordate; bracts much shorter than calyces	**15. R. humeana**
17b.	Dorsal petal broadly elliptic; bracts shorter than or equal to calyces	**16. R. forrestii**
18a.	Peduncle visible; dorsal petal obovate to obcordate	**14. R. cautleyoides**
18b.	Peduncle hidden; dorsal petal oblong to narrowly elliptic	19
19a.	Dorsal petal elliptic, 2.5–3.5 cm long; staminodes rhombic	**19. R. praecox**
19b.	Dorsal petal oblong, 1.5–2 cm long; staminodes narrowly obovate-cuneate	**20. R. kunmingensis**

8. TAXONOMY OF SPECIES

1. ROSCOEA ALPINA

When I was preparing a revision of the genus *Roscoea* (Cowley, 1982), much of the cultivated stock of this species at Kew and elsewhere was of unrecorded origin. The Chungtien-Lijiang-Dali Expedition to Yunnan in 1990 (CLD), the Baker, Burkitt, Miller and Shrestha (BBMS) expedition to Ganesh Himal, Nepal in 1992 and the Alpine Garden Society Expedition to China in 1994 (ACE) all changed this, and the collection at Kew has been much enhanced with new plants of known origin.

The plants portrayed here are two of the three collections of *Roscoea alpina* brought back from the BBMS expedition. The pale purple form (BBMS 2) was collected on 11 July 1992 above Gunga Bhanjyang at 3000 m, and the dark purple form (BBMS 3) on the same day in the same area, below Rupche Bhanjyang, at 3400 m. The former was growing in *Rhododendron* woodland, and the latter in a clearing in *Juniperus recurva* woodland. The third collection (BBMS 14), another pale form, was found on 16 July at Kal Khadga, above Tibling on the Ankhu Khola, at 2900 m on a dangerously precipitous open grassy slope on the edge of mixed, but mostly *Rhododendron*, woodland.

Roscoea alpina is a common species in the Himalayan range from Pakistan and Kashmir in the west to Bhutan in the east, at altitudes from 2000 m to 4300 m. It flowers at any time from the end of May to mid August. It was gathered on many occasions by British collectors in India at the end of the nineteenth century, and many specimens in the herbarium at Kew are around the hill stations in areas such as Simla, Mussooree, Sikkim and Kumaon. The most prolific collectors in those times were J.R. Drummond, J.S. Guthrie and T. Thomson; the more recent collections are from Nepal and have been made by collectors such as Sir Colville Barclay and P.M. Synge, A.D. Schilling, and C. Grey-Wilson and B. Phillips.

Many of the plants in cultivation that have until recently been referred to as 'alpina' in the nursery trade are, in fact, the pink form of *Roscoea scillifolia*. This species was formerly known from Western China, not an area where 'true' *R. alpina* occurs. The word 'formerly' is used here since *R. scillifolia* has not, unfortunately, been recollected on recent expeditions to the Lijiang area of Yunnan where Forrest, Handel-Mazzetti and Rock found plants at the beginning of the last century. The similarity between the two species is superficial, and when growing side by side, they cannot readily be confused. The inflorescence, and eventually the fruiting head, of *R. scillifolia* is held above the rosette of leaves on a short to long stalk (peduncle), and the small pink or deep purple flowers have narrowly elliptic, rather insignificant dorsal petals. The inflorescence of *R. alpina*, on the other hand, has no peduncle; only the flower and perianth tube are exserted from the calyx and the top of the leaves, and in the fruiting stage the capsules remain within the overlapping leaf sheaths which form the false stem. The significantly larger flowers have a rounded dorsal petal curved above the other flower parts, rather like a bonnet. A good photograph showing this illustrates Richard Wilford's article on *Roscoea* on page 68 of *Gardens Illustrated* for March, 2000.

40 | THE GENUS ROSCOEA
PLATE 1

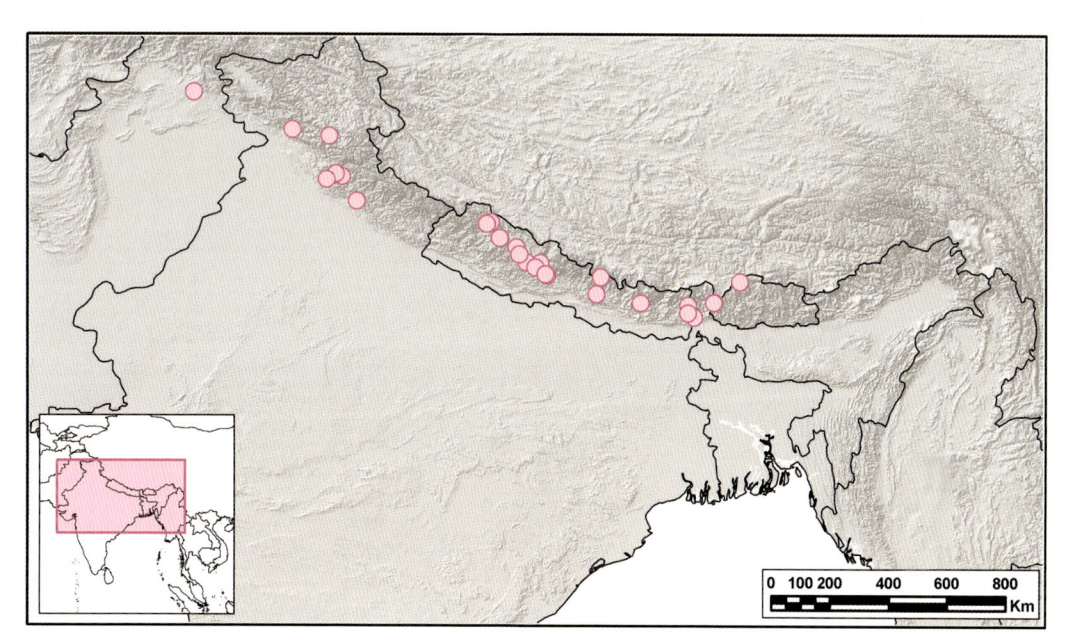

Map 4. Distribution of ***Roscoea alpina***.

Roscoea alpina has also been confused with *R. tibetica*, a species with a wide distribution from southeast Tibet and the Burma/China borders through to Sichuan and Yunnan in western China. The same comparisons could be made regarding the flower shapes as were made between the two species above. *Roscoea tibetica* and *R. scillifolia* have similarly shaped flowers, but those of *R. scillifolia* are significantly smaller than those of *R. tibetica* or *R. alpina*. *Roscoea alpina* and *R. tibetica* also differ in their leaf habit; *R. tibetica* normally remains compact, with its leaves arranged in a rosette, whereas at maturity the stem of *R. alpina*, with its more-or-less distichously arranged leaves, may be as much as 40 cm long. A significant difference, not immediately obvious unless the flowers are dissected, is the very short floral bract, no more than 1 cm long, at the base of the ovary in *R. alpina*; that of *R. tibetica* is 2 cm or more in length.

In more recent accounts, *Roscoea alpina* has been correctly identified, as in A.D. Schilling's account of his exploration of the Langtang valley of Nepal in 1969, published in the *Journal of the Royal Horticultural Society* volume 94 (5).

According to *The Garden*, 1914, *Roscoea alpina* was introduced into cultivation by Bees Ltd. saying that it was "a very hardy free-growing plant with fine purple flowers. Himalayas."

Both *Roscoea longifolia* and *R. intermedia* are conspecific with *R. alpina*. *Roscoea longifolia* was described from fruiting specimens in which the leaves were fully grown; the type collection came from very close to the type locality of *R. alpina*.

This species can be found in forest clearings or among low herbaceous plants in *Betula* woodland and *Rhododendron* or conifer forests, in open meadows in dry peaty soil or in short, damp grass. It can also be found on open south-facing hillsides up to the tree-line and in grassy patches on rock

Plate 1. ***Roscoea alpina*** forma ***pallida***, BBMS 2 (upper); forma ***alpina***, BBMS 3 (lower). Painted by Mark Fothergill, July 1995.

42 THE GENUS ROSCOEA
1. ROSCOEA ALPINA

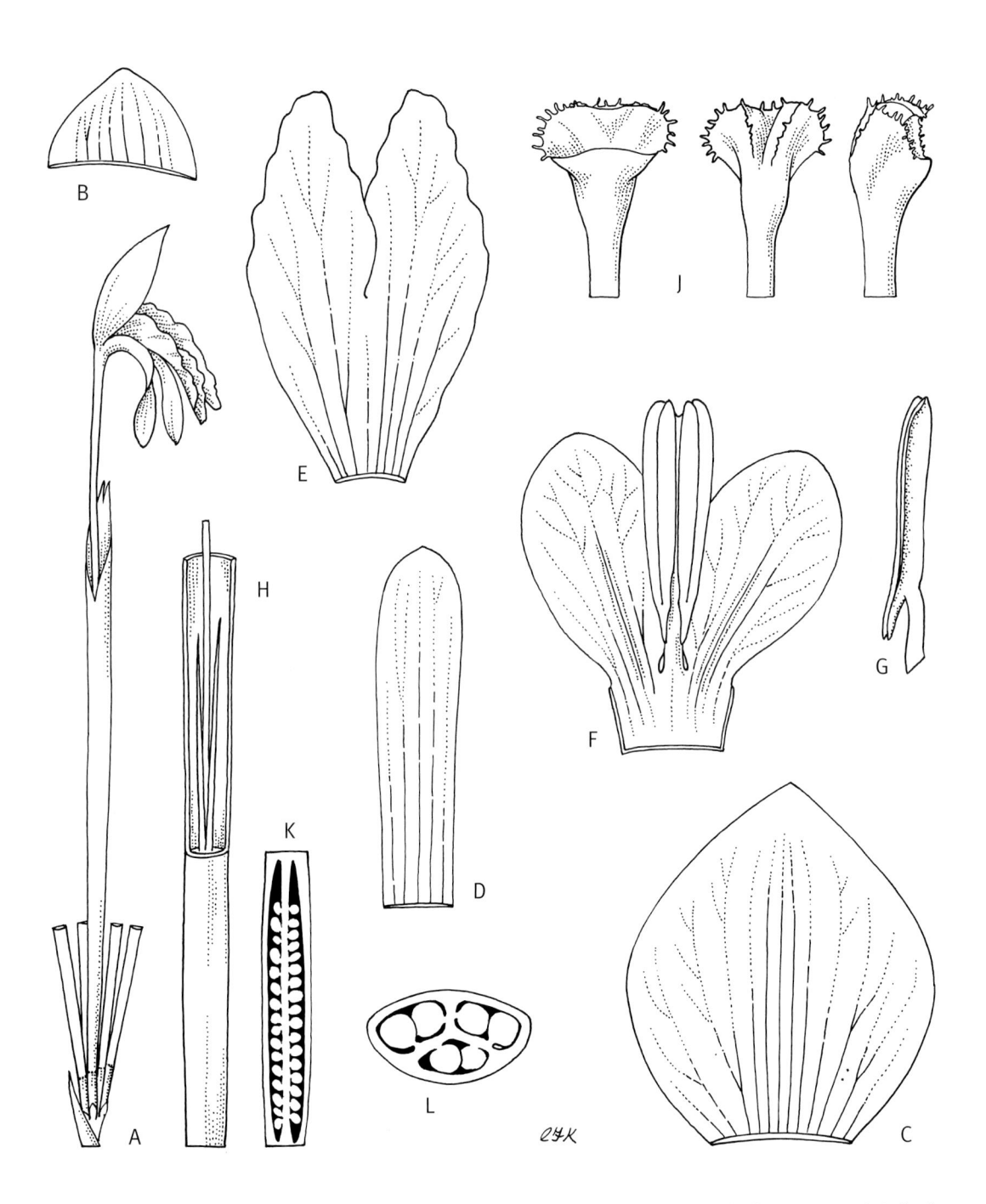

Fig. 14. Dissection diagram of ***Roscoea alpina*** by Christabel King. **A** inflorescence with 4 of 5 flowers removed; **B** floral bract; **C** dorsal petal; **D** lateral petal; **E** labellum; **F** stamen and staminodes; **G** stamen, side view; **H** ovary and base of style with epigynous glands; **J** stigma, 3 views; **K** ovary (longitudinal section); **L** ovary (transverse section).

face or ledges or on sides or beds of gorges at altitudes between 2000 and 4300 m. Plants may be found flowering from the end of May to mid-August.

From recent observations it would seem that this species is not at present under threat, as it is found over a wide area of the Himalayan range, but individual populations will inevitably be threatened by destruction of their habitat.

Roscoea alpina Royle, Ill. Bot. Himal. Mts.: 361, t. 89, fig.1a (1839). Type: India, Simla to Fagu, *Jacquemont* 1024 (isosyntypes K, LIV).
R. *purpurea sensu* Royle, Ill. Bot. Himal. Mts.: 361, t. 89, fig. 3 (1839).
R. *alpina* Royle var. *minor* Royle, Ill. Bot. Himal. Mts.: 361, t. 89, fig. 1b (1839). Type: not traced.
?R. *exilis* Horan., Prodr. Monogr. Scitam. 21 (1862). Type: *Jacquemont* exempl. 6.
R. *purpurea* Sm. var. *exilis* (Horan.) Baker in Hook. f., Fl. Brit. India 6: 208 (1890) as to type only.
R. *longifolia* Baker in Hook. f., Fl. Brit. India 6: 208 (1890). Type: Fagu, Simla, *Thomson* s.n. (holotype K).
R. *intermedia* Gagnep. in Bull. Soc. Bot. France 48: LXXIII (1901). Type: Northern India, Kulu (Plantes du pays des Koulos), *Ujfalvy* s.n. (holotype P).

ILLUSTRATION. *The Rock Garden* 26(4): fig. 115; text p. 323 (2000).

DESCRIPTION. *Rhizomatous, tuberous rooted herb* (6–)12–20(–40) cm tall. *Sheathing leaves* 2–3, obtuse. *Leaves* at flowering time usually 1–2 and under-developed, at maturity up to 6 and well-developed, 2.5–30 × 1.3–3.5 cm, linear, broadly elliptic or lanceolate, the first leaf slightly auriculate and widest at the base, the others widest at the middle, glabrous, sometimes hairy at the acute apex, mid-green above, paler below, side veins parallel; *ligule* area more or less semicircular, raised for 0.5 mm. *Inflorescence* more or less sessile, included within the sheathing leaves. *Bracts* shorter than the ovary, obtuse to somewhat truncate, 3–10 × 2–3 mm. *Calyx* 5.5–7.5 cm long, tubular to about 1 cm from the bluntly bidentate apex, much longer than its subtending bract, spotted. *Flowers* deep purple, mauve, lilac, pink or white, up to 5 per inflorescence, one open at a time; *perianth tube* usually white at base, purplish in the upper part, 5.5–7.8 cm long, usually exserted from the calyx by up to 3 cm; *dorsal petal* circular, cucullate, apiculate, (1.3–)1.6–1.8(–2.2) × (1.1–)1.5–1.7(–2.4) cm; *lateral petals* linear-oblong, (1.3–)1.5–1.6(–2.2) × (0.3–)0.5–0.6(–0.7) cm, obtuse; *labellum*

Fig. 15. **Roscoea alpina** forma **pallida**, BBMS 2, from Nepal. Cultivated at Kew and photographed by Andrew McRobb, R.B.G., Kew.

Fig. 16. ***Roscoea alpina***, cultivated at Kew. Photographed by Richard Wilford, R.B.G., Kew.

Fig. 17. ***Roscoea alpina***, growing in the Ganesh Himal, Nepal. Photographed by Bill Baker, 1992.

obovate, deeply bilobed, (1.4–)1.5–2(–2.6) × (1–)1.2–1.6(–2) cm, not deflexed, without an obvious claw, undulate at the margins. *Staminodes* circular to elliptic, 6–12 × 4–8 mm, more or less symmetrical, white, vein more or less median but not prominent. *Anther* white to cream, thecae 5–7 × 2–3.5 mm; connective elongation angled to the pointed appendages, together 1.5–2 mm long. *Ovary* 8–18 mm long, 2–3 mm in diameter. *Epignous glands* usually unequal, c. 6–8 mm long. *Capsule* 2.5–3.5 cm long, 5–8 mm in diameter. *Seeds* small, squarish in outline, dark brown, striate, slightly constricted near the middle, apiculate; aril shallowly lacerate.

Roscoea alpina Royle forma **pallida** Cowley, forma nov. a forma *alpina* floribus pallide nec saturate purpureis differt. Typus: Nepal, Ganesh Himal above Gunga Bhanjyang, 3000 m, *Baker, Burkitt, Miller & Shrestha* BBMS 2 (holotypus K).

DESCRIPTION. Differing from the typical form (forma *alpina*) by having pale purple flowers rather than mid-purple.

2. ROSCOEA PURPUREA

This species is the type of the genus, and also one of the most widespread and most variable in habit and form. Many of its various colour forms are now available in nurseries and there are numerous named cultivars. Until recently, a true red form seemed unlikely to occur, but thanks to the Baker, Burkitt, Miller & Shrestha expedition to the Ganesh region in Nepal in 1992, such a colour form is now known.

Roscoea purpurea occurs in the Himalayas from Himachal Pradesh in the west to the Assam/Bhutan frontier region in the east, and typically has pale purplish lilac flowers. There has been longstanding confusion between *R. purpurea* and the commonly cultivated darker purple-flowered species from Sikkim, *R. auriculata*. One of the causes of the confusion was the publication of an illustration (t. 4630) in *Curtis's Botanical Magazine* (Hooker, 1852) which appeared under the name *R. purpurea* but in fact depicts a species, *R. brandisii*, that is found only in the Khasia Hills in Assam. This is a smaller-flowered close relative of *R. auriculata*, and was first described as a variety of *R. purpurea* (Baker, 1894). When Schumann (1904) described *R. auriculata*, he also recognised the variety *brandisii* as a distinct full species. Of course, in 1852 *Roscoea auriculata* had not yet been described and so *R. purpurea* was the only published name to which the Botanical Magazine illustration could at that time be referred. The differences between the two species are discussed under *R. auriculata* (see p. 73 and under *R. tumjensis* p. 72).

The true *Roscoea purpurea* was mentioned in the first volume of the *Gardener's Chronicle* in 1841. It was flourishing outside at Edinburgh Botanic Garden, according to R. Graham and later at Kew, in 1889, according to Mr Nicholson. It had previously been thought of as a stove plant. From the 1930s true *R. purpurea*, as opposed to *R. auriculata* posing as *R. purpurea*, was being grown at Wisley and was known as *R. purpurea* var. *pallida*. *Roscoea purpurea* is again becoming more common in cultivation after recent introductions.

Several correctly named, botanically accurate paintings (listed below) of *Roscoea purpurea* had been published before 1852 and J.D. Hooker must have had access to these publications. It is therefore curious that he did not notice the differences between the plants depicted in the paintings and the Khasia and Sikkim specimens that he had written about and collected for Kew. In 1881, *R. auriculata*

THE GENUS ROSCOEA
PLATE 2

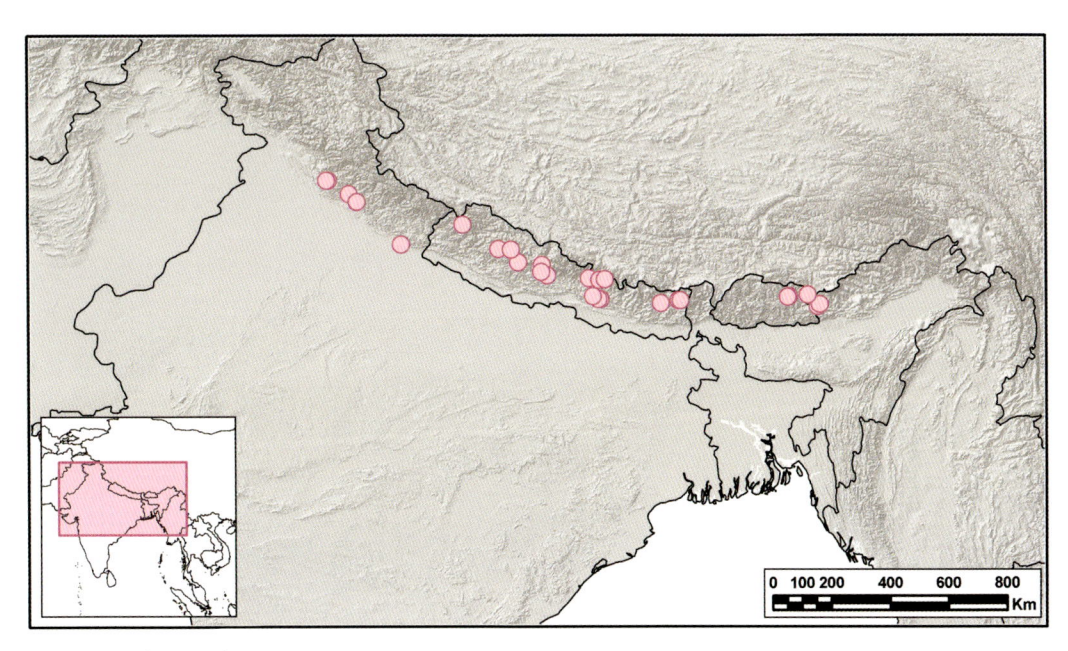

Map 5. Distribution of *Roscoea purpurea*.

was introduced from Sikkim by Henry John Elwes, who called it *R. purpurea* var. *sikkimensis*, although this name was never legitimately published. A grower, William Thompson of Ipswich, recognised the differences (Thompson, 1890) between Elwes's plant and the true *R. purpurea*. He described the differences in colour and habit and remarked that the specific epithet *purpurea* would be more appropriate for the Sikkim plant. In the same volume of the *Gardeners' Chronicle*, Series 3, Volume 8, there is a discussion of *R. auriculata* under the name *R. purpurea* var. *sikkimensis*; the accompanying illustration shows the strong auricles at the base of the leaves which are lacking in *R. purpurea*.

W.J. Hooker (1825) mentioned that the plant illustrated by J.E. Smith (1806), and used by him when describing *Roscoea purpurea*, was similar to the one he was depicting. He also remarked upon what he believed to be the inaccuracy of the artist when painting this illustration, because it appears too dark in colour, an unnatural indigo, although otherwise botanically correct. I have had the privilege of viewing the original painting in the Library at the Liverpool Museum, and I would describe the colour of the labellum as navy blue. I can only surmise that, if the artist was painting from a living specimen and knew that the flowers last only one day, he or she adjusted the colour after the plant had withered overnight. It is, however, more likely that he was trying to paint from drawings and dried specimens in Dr Buchanan's collection, and that he had been given insufficient information on the original flower colour. However, the painting does show the dorsal and lateral petals to be the familiar lilac-purple of that species; furthermore, the stem is painted as suffused with red, a state typical of some forms, both purple or red-flowered.

There is no doubt that the plant had originally been collected from the wild, and then cultivated at the Narainhetty Palace in Kathmandu in Nepal, and had been sent from there to England by Dr

Plate 2. *Roscoea purpurea* forma *purpurea*. BBMS 41. Painted by Christabel King, July 1999.

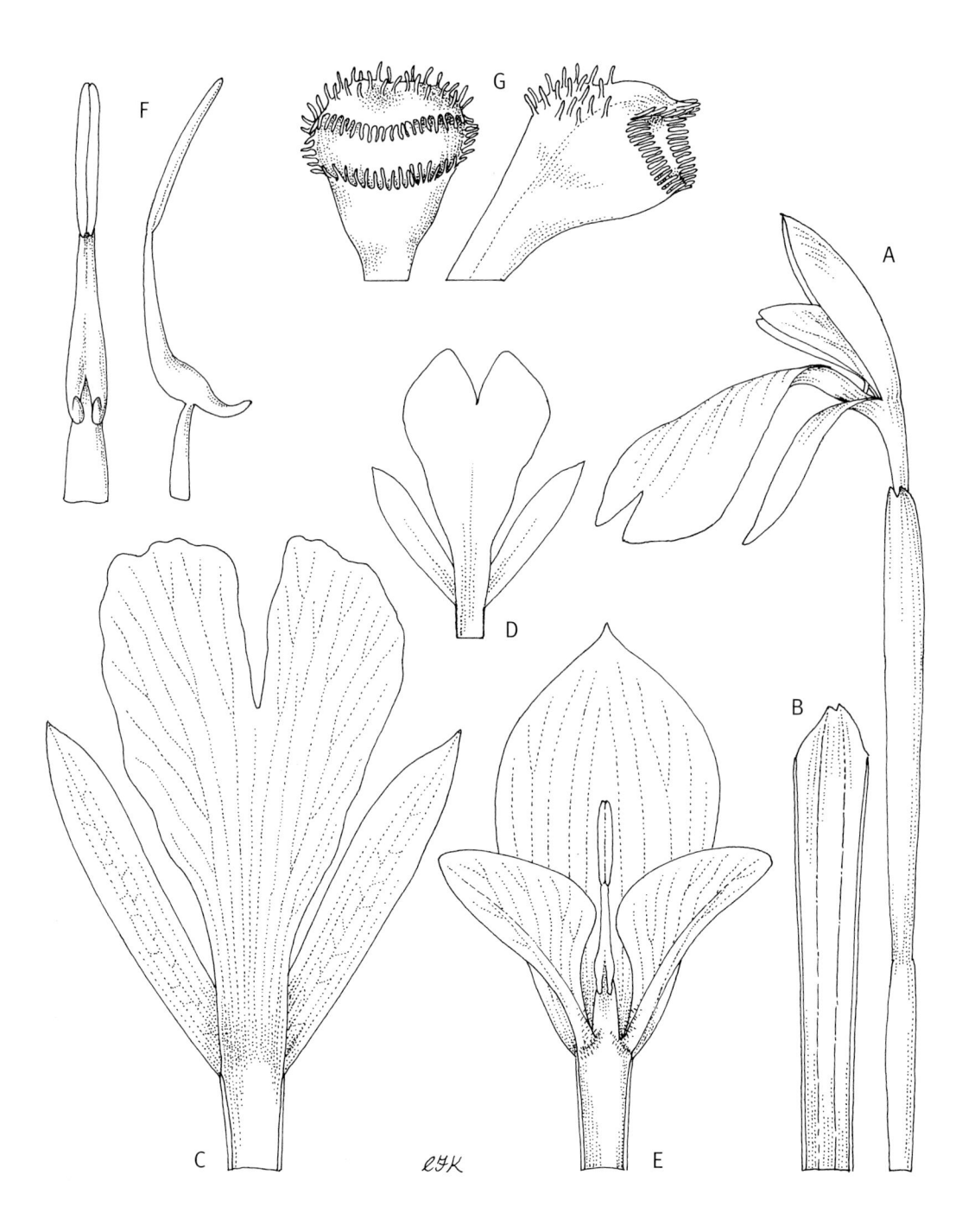

Fig. 18. Dissection diagram of **Roscoea purpurea** by Christabel King. **A** flower; **B** inner bract; **C** lower portion of corolla; **D** lower portion of corolla from dried specimen BBMS 41; **E** upper portion of corolla; **F** stamen, two views; **G** stigma, two views.

Buchanan. The first living specimen of the genus named after him that Roscoe was to see, was sent to him at the Botanic Garden in Liverpool by Dr Wallich of Calcutta; in 1828 he wrote that it had flowered in 1821, some years after the species had been described by J.E. Smith. Herbarium specimens taken from plants cultivated at Liverpool at this time have been annotated 'Roscoea speciosa' Aug. 21 [presumably this means August 1821 and not August 21st], but the unknown writer thought that the plant was different from Smith's one, and cited Hooker's *Exotic Flora* t. 144, which accurately illustrated *Roscoea purpurea*.

For many years a form of *Roscoea purpurea* with large white flowers streaked with purple was wrongly known as 'R. procera'. "Procera" was the specific epithet given below the illustration (t. 242), painted by Gorochaud that accompanied Nathaniel Wallich's description of 'R. purpurea gigantea' in his *Plantae Asiaticae Rariores*, part 3, in 1832. It is not clear why the description and illustration were published under different names. The painting is splendid, and is of a typical pale purple *R. purpurea* which was, however, according to the text, a very tall form. Some forms of *R. purpurea* at the western and eastern extremes of its range are paler and larger-flowered than those forms found in the central Himalayan districts. However, completely white-flowered forms are commonest in eastern Nepal.

The plant which later came into cultivation and which was called 'procera' was a Kingdon Ward collection from Assam, but the same form can also be found in Bhutan. Kingdon Ward's field notes on his collection no. 13755 state: 'KW 13755 Roscoea purpurea? Height 4–6". The leafy stem bears a succession of large yawning flowers, white splashed purple on the lower lip. Forms large colonies on steep rocky pine covered slopes high above Dirang Dzong. 7000–8000 ft.'. July 1938'. Strangely, the specimen of this number at the Natural History Museum bears slightly different field notes: 'Assam, Orka La, 8000–9000', June 1938, Kingdon Ward 13755'. Whatever the correct collecting data may be, the subsequent history of Kingdon Ward's plant is recorded in 1948 in the *Gardener's Chronicle*. F.C. Wood of Worthing wrote that "This species, if species it can be called — was sent home under the number KW 13755. Plants raised by Colonel (F. C.) Stern (OBE, MC), at Highdown flowered well and the new introduction was granted an A.M. (Award of Merit) on July 2 1946. There is apparently some doubt in the mind of botanists as to whether *R. procera* merits specific rank or whether it should be regarded as a variety of *R. purpurea*." He then goes on to discuss J.M. Cowan's views in his *Review of the genus Roscoea* which had been published in 1938, the same year in which the

Fig. 19. **Roscoea purpurea** forma **purpurea**, Ganesh Himal, Nepal. Photographed by Bill Baker, 1992.

Fig. 20. ***Roscoea purpurea*** forma ***purpurea***, SW of Amjilassa, Ghunsa Khola, Nepal, 2500 m, KEKE 258. Photographed by R.B.G., Kew.

plant had been collected in Assam. Cowan was therefore unaware of the white-flowered plant, and viewed *R. procera* as being the correct name for the true *R. purpurea*. Cowan also stated that *R. procera* and *R. purpurea* were very similar. He also seems to have thought that the name *R. purpurea* really belonged to what we now know as *R. auriculata*, because he described *R. purpurea* as having auricled leaves, dark purple flowers and white staminodes, thus demonstrating the continued confusion. The *Journal of the Royal Horticultural Society* in 1946 recorded the Award of Merit and, after a short description of the plant, stated "The species is figured in the Botanical Register of 1840 (vol. 26, t. 61) as *R. purpurea*" — this statement is actually correct! The white-flowered form is described formally as *R. purpurea* forma *alba* below.

The exciting new red-coloured form of *Roscoea purpurea* was exhibited by Kew at one of the Royal Horticultural Shows at Vincent Square on 16 August 1994, and awarded a Preliminary Commendation. It was collected several times from the same place, and BBMS 43 was chosen as the type collection for the cultivar name 'Red Gurkha'. According to the rules, the only collection which can be called by that name is the collection chosen at the time of publication. All collections of red forms are described below as *R. purpurea* forma *rubra*. The red form of *R. purpurea* was found in the valley of the Buri Gandaki, a vigorous river that divides the two high ranges of the Gorkha Himal and the Ganesh Himal. It is known from a single locality at an approximate altitude of 1900 m, the plants were scattered amongst open scrub and on steep terrace walls at the edge of a village. Such disturbed and open habitats are favoured by the species of *Roscoea* that were encountered during that expedition. However, no other forms of *R. purpurea* occur in the immediate vicinity of *R. purpurea* forma *rubra*, and the uniformity of flower colour within the population suggests that the form would come true from seed. The plants included ones with the leaf sheaths suffused with red, and forms without the red suffusion. Forms of this kind can also be seen in the purple-flowered form. Plants with shorter stems occurred in drier, more exposed areas and taller specimens were found in lush grass. On the whole, red-flowered plants tended to be smaller than ones of the typical form. The restriction of the red-flowered form to a single locality means that its conservation status must be regarded as vulnerable.

The two illustrations (Plates 2 and 3) both depict plants collected on the 1992 expedition. The purple form is BBMS 41, collected on 1 August in the Ganesh Himal at Abuthum Lekh near Yarsa at 2700 m; it grew in shallow soil pockets on top of boulders in grazed meadow banks. The dense population contained white, pink and purple colour forms. The red form is the collection chosen for the type of forma *rubra* (see below).

Little has been written of any medicinal properties of species of *Roscoea*, but it has been noted that the roots of *R. purpurea* are used in veterinary preparations.

Roscoea purpurea can be found in both damp and dry positions, in grassland on alpine slopes, in thin soil over rocks, on moist rock faces, on terrace walls, in exposed south-facing positions, in thick herbaceous growth of grasses and shrubs, in clearings and in shade at edges of forests and woodland. It has been collected between 1520 and 3100 m a.s.l., and flowers from the end of June to September.

Fig. 21 (left). **Roscoea purpurea** forma **purpurea** collected in Nepal by Brian Halliwell. Cultivated and photographed at Kew by Richard Wilford, R.B.G., Kew.

Fig. 22 (right). Red stemmed form of **Roscoea purpurea** forma **purpurea**. Cultivated and photographed at Kew by Richard Wilford, R.B.G., Kew.

2. ROSCOEA PURPUREA

NOTE. The exact publication date of *Roscoea purpurea* (and consequently the genus *Roscoea*), remains uncertain. The second volume of J.E. Smith's *Exotic Botany* bears a date of 1806, but within it are plates dated 1805 to 1808. The plates may originally have been published separately, and put together as a volume later. Roscoe believed that it had been published in 1805 and he cited that date in his *Monandrian Plants* (1828). Here I use the date 1806.

Roscoea purpurea Sm., in Exotic Botany 2: 97 t. 108 (1806). Type: Nepal, Narainhetty, *Buchanan* s. n. (holotype LINN, isotypes BM, LIV).

R. *procera* Wall., Pl. Asiat. Rar. 3: t. 242 (1832).

R. *purpurea* [var.] *gigantea* Wall., Pl. Asiat. Rar. 3: 22 (1832), figd. on t. 242 of the same work as R. *procera* Wall. (See text above).

[R.*purpurea* Sm. var. *exilis sensu* Baker in Hook. f., Fl. Brit. Ind. 6: 208 (1890), excl. type].

R. *purpurea* Sm. var. *procera* (Wall.) Baker in Hook. f., Fl. Brit. Ind. 6: 208 (1890).

[R. *exilis sensu* K. Schum. in Engl., Pflanzenr. 4 (46): 119 (1904) partly — as to the *Wallich* specimen only].

R. *purpurea* var. *pallida* Hort.

R. *purpurea* var. *purpurea* Hara in Hara *et al.*, Enum. Fl. Pl. Nepal 1: 61 (1978) excl. *Stainton 3833*.

[R. *auriculata sensu* Hara in Hara *et al.*, Enum. Fl. Pl. Nepal 1: 61 (1978), partly — as to *Kanai, Hara & Ohba 723607*, cited as TI723601].

ILLUSTRATION. Hooker's *Exotic Flora* 2: fol. & t. 144 (1825); Loddiges's *Botanical Cabinet* 15: t. 1404 (1828); Roscoe's *Monandrian Plants of the Order Scitamineae* 2, t. 64 (1828); Lindley's *Botanical Register* 26: t. 61 (1840).

There is a good photograph of Tony Schilling's collection (no. 1185) from Kathmandu valley taken in Roy Lancaster's garden on page 33 of the latter's excellent book 'A Plantsman in Nepal'.

DESCRIPTION. *Perennial herb* (12–)25–38(–55) cm tall. *Sheathing leaves* 0–2, obtuse to truncate, cleft. *Leaves* 4–8, elliptic, lanceolate to oblong-ovate, sometimes falcate, apex acuminate and sometimes ciliate, slightly auriculate at base on lower leaves, side veins parallel, (4–)14–20(–25) × (0.8–)1.5–4.7 (–5.5) cm. *Leaf sheaths* reticulately veined, sometimes strongly diffused with purple or red. *Ligule* area ± semi-circular, ligule raised for 1 mm. *Inflorescence* enclosed in upper leaf sheaths, only upper part of bracts and flowers exserted. *Flowers* light purple, mauve, lilac, pink, red, white or white with purple markings, usually only one flower open at a time. *Bracts* pale green, shorter or longer than the calyx, acute, 7–13 × 0.5–2 cm. *Calyx* sharply bidentate, apiculate, pale green sometimes marked pink, 5–8.8 cm long. *Corolla tube* hardly exserted from calyx, mauve or white, 6.5–10 cm. *Dorsal petal* narrowly elliptic, hooded, apiculate, 3–6 × 1–2.8 cm. *Lateral petals* linear-oblong 3.3–6.5 × 0.3–1.2 cm. *Labellum* clawed, obovate, angular-obovate, very broadly obovate or very broadly truncate-obovate, entire or shallowly to deeply lobed, 3.5–6.5 × 2–5 cm including the 0.7–1.2 cm claw. *Staminodes* obliquely spathulate, white, red or mauve-veined white, 2–4 (including long, narrow claw) × 0.6–1.1 cm. *Anther* white, thecae 7–15 × 1.5–2 mm; connective elongation angled to the creamy-yellow, pointed appendages, together 9–25 × 2–3 mm. *Epigynous glands* 7.5–9.5 mm long. *Style* and *stigma* white. *Ovary* triangular in section, 2–3.8 × 0.3–0.5 cm. *Seeds* elliptic to triangular, aril shallowly lacerate.

Plate 3. *Roscoea purpurea* forma *rubra*. BBMS 45. Painted by Christabel King.

THE GENUS ROSCOEA | 53
PLATE 3

2. ROSCOEA PURPUREA

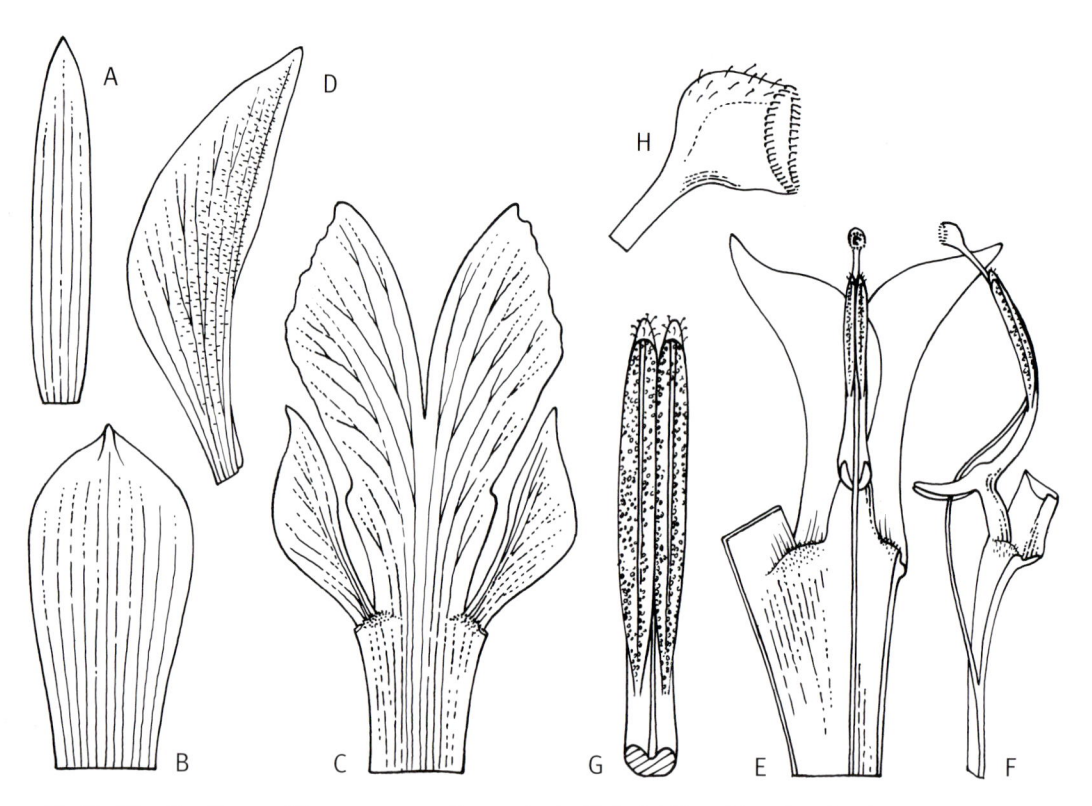

Fig. 23. Dissection diagram of **Roscoea purpurea** forma **rubra** by Christabel King. **A** lateral petal; **B** dorsal petal; **C** labellum and staminodes; **D** staminode; **E** stamen, style and staminodes, front view; **F** stamen and style, side view; **G** anther; **H** stigma.

Fig. 24 (left). **Roscoea purpurea** forma **rubra**. Cultivated in the Alpine House at Kew and photographed by Richard Wilford, R.B.G., Kew.

Fig. 25 (right). **Roscoea purpurea** forma **rubra** 'Red Gurkha'. Collected, cultivated and photographed by Bill Baker.

Two colour forms are recognised and described here:

Roscoea purpurea Sm. forma **alba** Cowley, forma nov. a forma *purpurea* floribus albis labello purpureo-maculato nec omnino pallide purpureis neque rubris differt. Typus: Assam, Orka La, 2440–2740 m, June 1938, *Kingdon Ward* 13755 (holotypus BM).

DESCRIPTION. Differs from the typical form (forma *purpurea*) in having flowers which are white with a labellum splashed deep purple rather than wholly pale purple or reddish.

Roscoea purpurea Sm. forma **rubra** Cowley, forma nov. a forma *purpurea* atque forma *alba* floribus utrinque rubris nec pallide purpureis neque albis purpureo-maculatis differt. Typus: Plant cultivated at Kew, September 1993 (Kew accession number 1992-2314) from material from Nepal: Buri Gandaki valley, 1900 m, 1 August 1992, *Baker, Burkitt, Miller & Shrestha* (BBMS) 45; (holotypus K).

DESCRIPTION. Differing from the typical form (forma *purpurea*) and from forma *alba* by having flowers in the red range.

Fig. 26. **Roscoea purpurea** forma **alba**. Cultivated at W. Ingwersen's nursery. Photographed by Brian Mathew, 1958.

3. ROSCOEA CAPITATA

Plants of this Himalayan species of *Roscoea*, collected in Nepal in 1970 (as BH 34) by Brian Halliwell, a former Assistant Curator in the Alpine and Herbaceous Section at Kew, were cultivated at Kew but unfortunately lost after a few years. However, the 1992 Oxford University Ganesh Himal expedition led by William Baker, which was instrumental in the introduction of several interesting plants, was successful in finding *R. capitata* again and, to date, the resulting plants are thriving in cultivation.

Roscoea capitata is the only Nepalese species of the genus which holds its inflorescence above the leaves on a peduncle, and in the field it is hard to confuse with any other species from the area. William Baker states that his team's first discovery of the species was 'in the airy rubble of the terrace walls, often leaning out of vertical, grassy banks', where it was common. It is this early collection (*Baker, Burkitt, Miller & Shrestha* 13), made in mid-July, from Tibling, Ankhu Khola, that is illustrated here. They found the species again on 7 August, also on disturbed ground, near Pansing in the valley of the Chhuning Khola, Buri Gandaki, where scrub had been cleared; two collections were made in this locality (BBMS 55 & 56).

3. ROSCOEA CAPITATA

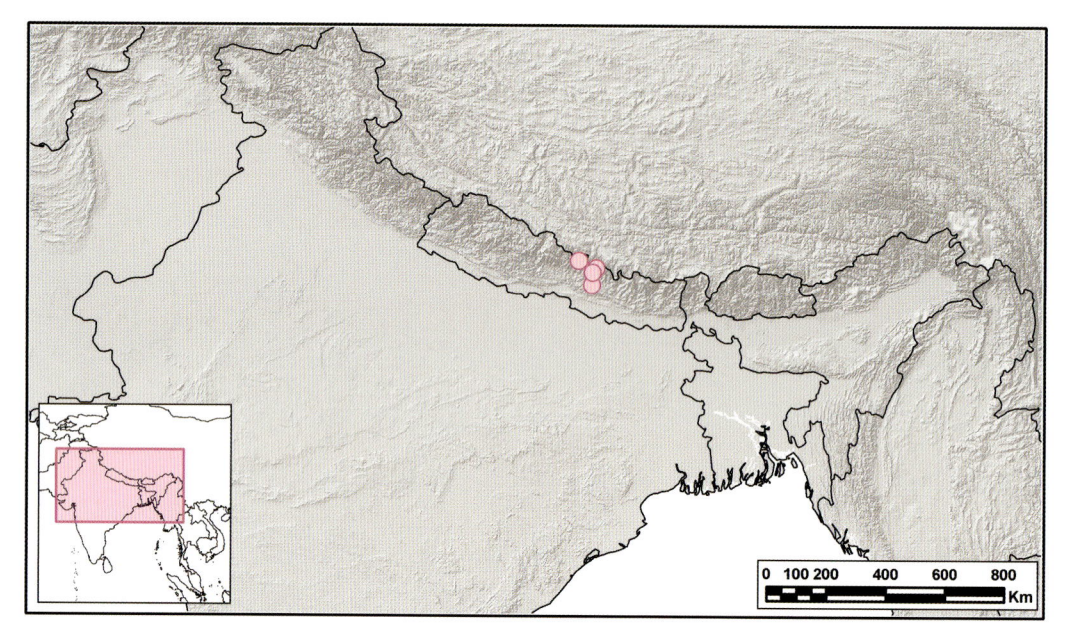

Map 6. Distribution of ***Roscoea capitata***.

It is distinct from other species of *Roscoea* from the area in its long, narrow, undulate leaves which have no marked distinction between the sheathing part and the lamina, which is held at an angle to the stem. The cone-shaped cluster of bracts, which become recurved at their tips, is positioned at the end of a distinct peduncle. The comparatively small flowers appear in succession from between the bracts; often more than one flower is open at a time in a single inflorescence. Their colour has been described variously as magenta, pale to deep purple, mauve, blue, pink or white; the plants of BBMS 13 cultivated at Kew have lilac purple or dark purple flowers. The calyces are distinctly ciliate-hairy on the veins and margins and are just hidden by the slightly longer bracts until flowering when they become shortly exserted. Another characteristic is the unusually long sickle-shaped staminodes, which are almost the same length as the dorsal petal, recalling the Chinese species, *R. schneideriana*. The colour of the staminodes is variable; in BBMS 13, illustrated here, they are predominantly white, but are purple at the apex, on the vein and at the margins, whereas in BBMS 55 they are entirely purple.

Gagnepain (1902) described two varieties of *Roscoea capitata*, both from China, but both, in my opinion, belong to other species. They are similar to *R. capitata* in having pedunculate inflorescences, but differ in other respects. *R. capitata* var. *purpurata* is a synonym of *R. cautleyoides* forma *sinopurpurea* and *R. capitata* var. *scillifolia* is now known as *R. scillifolia*. The pale pink form of the latter is often erroneously called 'R. alpina' in the horticultural trade; the true *R. alpina* is a distinct species from the Himalaya.

In Britain, the majority of herbarium collections from Nepal are housed at the Natural History Museum, London (BM) and this is where many of the modern collections of *Roscoea capitata* can

Plate 4. ***Roscoea capitata.*** BBMS 13. Painted by Christabel King, July 1997.

THE GENUS ROSCOEA
PLATE 4

58 THE GENUS ROSCOEA
3. ROSCOEA CAPITATA

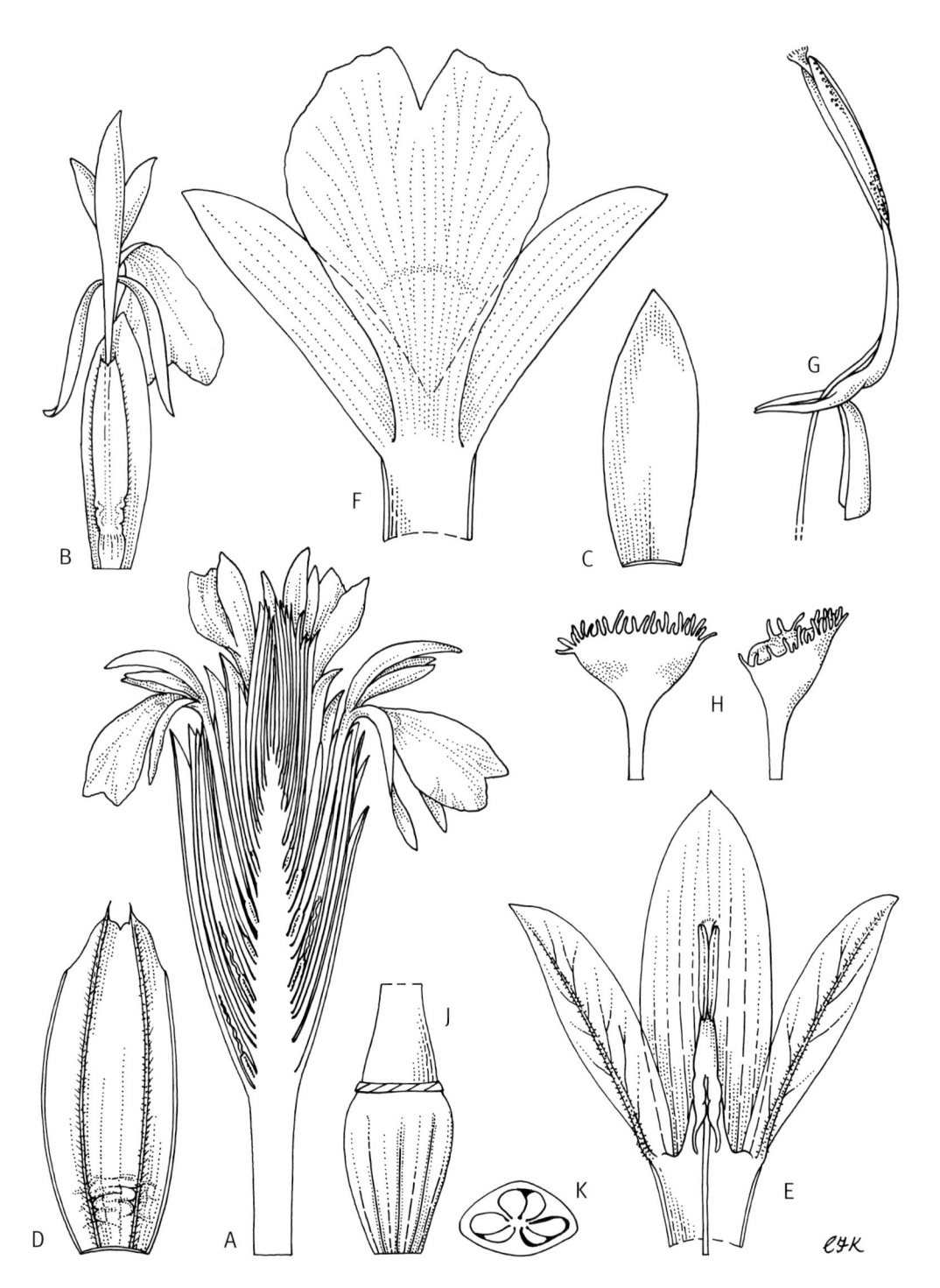

Fig. 27. Dissection diagram of ***Roscoea capitata*** by Christabel King. **A** inflorescence (transverse section); **B** flower and bract abaxial view; **C** bract; **D** calyx opened out; **E** upper portion of flower and stamen; **F** lower portion of flower and stamen; **G** stamens and upper part of style; **H** stigmatic surface, abaxial view and side view; **J** ovary; **K** transverse section of ovary.

be found. Collectors such as Gardner, Hara *et al.*, Polunin & Stainton brought back many specimens from their expeditions to Nepal in the middle years of the last century. These show the rather restricted distribution of the species. The centre of distribution seems to be around Trisuli; there are collections from Shiar Khola (*Gardner* 847), Langtang (*Polunin* 641 and *Syms* s.n.), Trisuli Khola (*Kanai, Hara & Ohba* 723600), Ankhu Khola (*Stainton* 3833), Kapur Gang (*Bailey's collector* s.n.), as well as the *Halliwell* collection from near Syarpagaon in 1970. Hara (1978) was mistaken in identifying *Stainton* 3833 as *R. purpurea* Sm. var. *purpurea*.

Two of the three type sheets in the main herbarium at Kew are wrongly numbered as *Wallich* 6239 and 6329; all should be numbered 6529. A search in the Herbarium of the Honourable East India Company (the Wallich Herbarium) at Kew revealed that the other two numbers belong to quite different taxa, neither of them a ginger.

Dr John Macqueen Cowan, who worked for the Indian Forest service as a Superintendant at the Royal Botanic Garden at Sibpur, Calcutta, reported that '*Roscoea capitata* grows in damp meadows on the fringes of forests and has blue-purple flowers.....clusters of fleshy rhizomes like spindles and is used in China as a cure for aches, indigestion, belch and various aches in the abdomen.'

Roscoea capitata is confined to an area to the north-west of Kathmandu in Central Nepal. It is locally common on open grassy hillsides or in damp gulleys and on stony slopes or loose stone walls. It may be an early coloniser in vegetation succession on disturbed ground around villages. It grows at altitudes between 1200 and 2600 m, and flowers from June to September.

Fig. 28. *Roscoea capitata*. Cultivated in the Alpine house at Kew. Photographed by Richard Wilford, R.B.G., Kew.

Roscoea capitata Sm. in Trans. Linn. Soc. London 13: 461 (1822). Type: Nepal, *Wallich* 6529 (holotype K; isotypes BM, CGE, E, G, K).

ILLUSTRATION. Wallich: *Plantae Asiaticae Rariores* 3: 35, plate 255 (1832); Horaninow: *Prodromus Monographiae Scitaminearum*: 20 (1862).

A photograph of this species can be found in William Baker's first account of his Ganesh Himal expedition in the *Alpine Garden Society Bulletin*, March 1994.

DESCRIPTION. *Rhizomatous, tuberous rooted herb* to 45 cm tall. *Sheathing leaves* 1–3, obtuse, soon splitting. *Leaves* 3–9, linear, rarely lanceolate, 6–34 × 0.9–4.5 cm, angled to the stem, narrowed at the base, bases covering stem, acute to acuminate, glabrous or ciliate on margins and keel, mid-green, finely punctate above, paler below, main veins prominent, side veins diverging. *Ligule* area ± semi-circular, ligule raised for 0.5–9.5 mm, pinkish, sometimes interrupted or bilobed at the keel. *Inflorescence* 3–6.5 × 0.5–3.5 cm, pedunculate, peduncle exserted 3–11 cm above base of topmost leaves. *Bracts* narrowly ovate, longer, equal to or slightly shorter than the calyx depending on maturity, acute, ciliate, green; lowest bract tubular, 3–4 × 0.5 cm, soon splitting to the base. *Calyx* 2.5–3.5 cm long, shortly but sharply bidentate, hairy at apex, veins and membranous margin conspicuously ciliate. *Ovary* 5–6 × 2.5 mm, glabrous, pink. *Flowers* magenta, purple, mauve, pink or white, few to many, in older plants several opening at the same time. *Corolla tube* c. 3.5 cm long, exserted no more than 1 cm above the bract and calyx. *Dorsal petal* elliptic to ovate, 1.5–2.6 × 0.8–1.2 cm, without basal claw, cucullate, apiculate. *Lateral petals* linear-oblong, 1.9–2.6 × 0.4–0.7 cm. *Labellum* obovate, 2–3 × 0.8–2 cm, including the 4–5 × 9 mm claw, sometimes deflexed, lobed for about a quarter of its length, each lobe conspicuously emarginate. *Staminodes* obliquely spathulate, 2–2.2 cm long, including 8 mm claw, 4.5–8 mm broad, thickened vein excentric, papillose. *Filament* 3–4 × 3 mm, creamy-white, streaked purple. *Anther* creamy white, thecae 5–7 × 1.5 mm, connective elongation angled to the spreading pointed appendages, swollen at junction with connective, together 10–14 × 2 mm. *Epigynous glands* 4–5 mm long. *Capsule* clavate, 2.2–2.5 cm long. Seeds not seen.

4. ROSCOEA GANESHENSIS

At first glance one might easily take this plant to be a rather stunted form of *Roscoea purpurea*, coming as it does, from a locality where the latter species is quite common. However, there are many significant differences, and these led to the description of this taxon as a new species in 1996.

Roscoea ganeshensis was discovered by Bill Baker, Tom Burkitt, Jonathan Miller and Rhidaya Shrestha on 2 August 1992 during the Oxford University Botanical Expedition to the Ganesh Himal in central Nepal. Two collections were made at the same place (BBMS 34 and BBMS 50). A striking red form of *R. purpurea* (forma *rubra*), had been found the previous day, so the appearance of a second new *Roscoea* was a remarkable piece of good fortune. Had they unwittingly strolled into the very centre of *Roscoea* diversity, the team asked themselves?

The genus is indeed very diverse in this area. *Roscoea purpurea* is abundant in disturbed areas and rough pasture from about 1500 m upwards; *R. alpina* is scattered in damp woodland, both in the shade of trees and in clearings, at approximately 2500 m and above. *Roscoea capitata* is thought to be restricted to central Nepal, and on the same expedition it was found to be locally common around paths and in villages at approximately 2000 m. The little-known species *R. tumjensis* also occurs in the Ganesh Himal and was found by the same expedition team in that area.

The discovery of *Roscoea ganeshensis* thus brings the total number of *Roscoea* species from this area to five. The type collection was made in the spectacular valley of the Buri Gandaki at about 1900 m where a single colony of around 200 individuals was found on a steep bank alongside a path. The plants were growing in loose, rocky soil with grasses, ferns, selaginellas and orchids in a fairly disturbed site of the kind that seemed to be favoured by most of the *Roscoea* species seen during the expedition. As soon as *R. purpurea* was found at the same site as *R. ganeshensis*, and the

Plate 5. *Roscoea ganeshensis.* BBMS 34. Painted by Christabel King, September 1993.

THE GENUS ROSCOEA | 61
PLATE 5

4. ROSCOEA GANESHENSIS

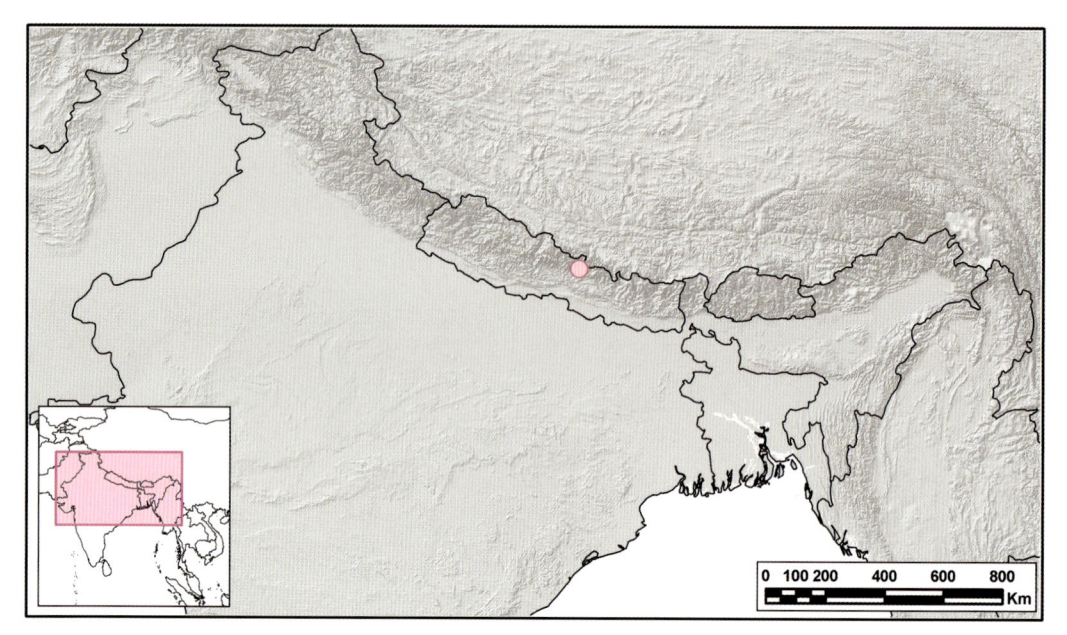

Map 7. Distribution of ***Roscoea ganeshensis***.

two species compared, the team became more certain that the plants now named as *R. ganeshensis* were really quite distinct. The initial judgement of the expedition members was that this new species was 'a more tidy plant than typical *R. purpurea*'.

Roscoea ganeshensis is readily identified in the field by its very short internodes, finely hairy leaves and the strongly shouldered appearance of the crumpled labellum, which is held above and at right angles to the claw when the flower first opens. The hooded dorsal petal is also somewhat paler than the labellum. This tends to be emphasised by the presence of a dark purple marking in the throat, but this character is, unfortunately, not constant, at least in cultivation. The bracts and the calyx are also hairy like the leaves. Although this species may not have the instant horticultural appeal of its extravagant neighbour, *R. purpurea* forma *rubra*, it is a plant of unparalleled, subtle charm when seen *en masse* in its natural surroundings. The type locality is, to date, the only one known.

In cultivation, the immature flower buds are totally enclosed by a tubular calyx. However, during development, the flower breaks through one of the two hyaline keels, so that the calyx apex at first remains completely tubular, only eventually splitting as far as the tip when the flower is fully open.

Roscoea purpurea has a dorsal petal which curves towards the lower parts of the flower, differing from *R. ganeshensis*, in which the dorsal petal is held erect.

The collection BBMS 50 of *Roscoea ganeshensis* was shown by Kew at one of the Royal Horticultural Society's Shows in London in 1999, and was awarded a Preliminary Commendation (P.C.) (Fig. 31). The report in the *Quarterly Bulletin of the Alpine Garden Society* in June 2000 commented that the collection shown was 'much more handsome and more uniformly rich purple than that illustrated (BBMS 34) in the Botanical Magazine'.

The limited data suggest that this species flowers during August and September.

THE GENUS ROSCOEA
4. ROSCOEA GANESHENSIS

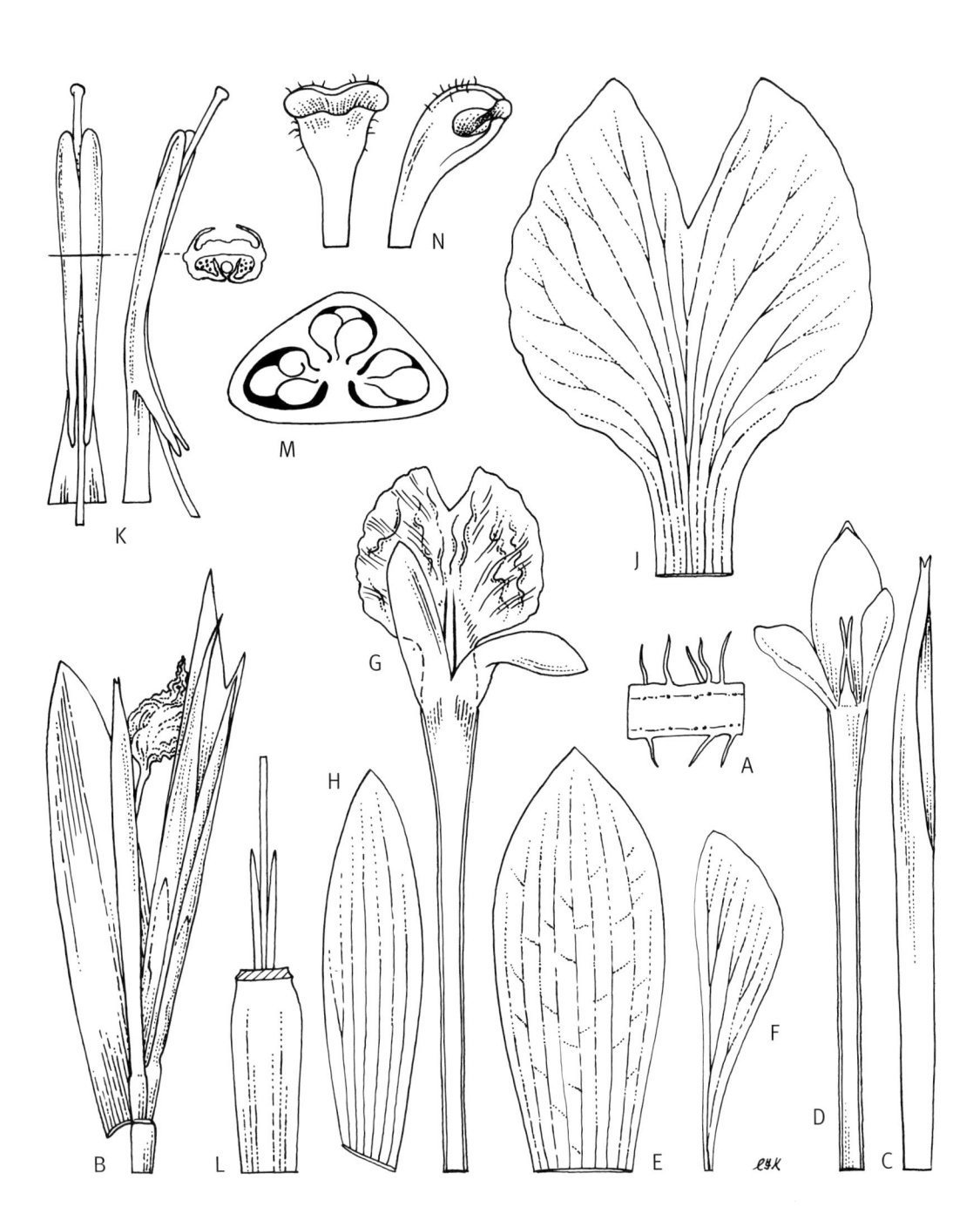

Fig. 29. Dissection diagram of ***Roscoea ganeshensis*** by Christabel King. **A** section of leaf lamina; **B** bract, with dead flower & two buds; **C** calyx; **D** dorsal part of flower, abaxial view; **E** dorsal petal; **F** staminode; **G** ventral part of flower, adaxial view; **H** lateral petal; **J** labellum; **K** stamen and upper part of style, with transverse section; **L** ovary, base of style and epigynous glands; **M** ovary, transverse section; **N** stigma, front and side views.

4. ROSCOEA GANESHENSIS

Roscoea ganeshensis Cowley & W.J. Baker in Curtis's Bot. Mag. 13(1): 10-13 (1996). Type: Nepal, Buri Gandaki valley, near Abuthum Lekh, 1900 m, 2 August 1992, *Baker, Burkitt, Miller & Shrestha* 34 (holotype K).

DESCRIPTION. *Perennial herb* 12–15 cm tall. *Rhizome* short, giving rise to swollen tuberous roots to 5.5 cm long, narrowing to fibrous, contractile roots with white tips, to 8–10 cm long. *Prophyll* a bladeless sheath, transparent-whitish with brown veins, tubular, apex obtuse, 6 mm long. *Stem* formed by overlapping leaf sheaths, internodes very short, compressed, elliptic in cross section. *Leaf sheaths* mid yellow-green, tubular, hyaline-keeled, contiguous on to the back of leaf blade as mid vein, soon splitting along hyaline line opposite leaf blade so that eventually only tubular at base, strongly rugose, with darker green reticulations, 2.5–6.5 × 1.1–2 cm. *Ligule* at base of leaf blade transparent, semi-circular, truncate, continuing along upper edge of leaf sheath, 0.5–2 mm high. *Leaves* 5–6, distichous, falcate, congested, held at right angles to the stem (apart from unkeeled leaf below inflorescence which is held erect), ovate to ovate-lanceolate, 6–19 × 2.7–5.3 cm, mid yellow-green, paler beneath, shortly, densely pubescent on both surfaces; bases sometimes slightly auriculate; edges undulate; apex shortly acuminate; underside of leaf with well-defined midrib 0.5–1 mm thick. *Bracts* green, elliptic, tubular, soon splitting on emergence of the flowers, outer surface shortly pilose; first or inflorescence bract reticulate, hairy, tubular to within 2.3–3.7 cm of apex, gradually narrowing to acuminate apex, 8–13 × 1–2 cm; floral bracts each supporting one flower, mid yellow-green, reticulately veined, narrowing to acute apex, 6–10.5 × 1.1–2 cm when mature and opened out. *Epigynous glands* 2, 4 mm long when immature, 7–8 mm long when mature. *Ovary* trilocular, triangular to semicircular in section, 5–11 × 4–4.5 mm, ovules many. *Calyx* greenish-white, paler at apex, spotted, totally enclosing flower at first, 2-keeled, sparsely pubescent to ciliate on veins and keels, 4.7–8.3 × 0.85–5 mm, tubular but split below the apex for 1–2 cm when mature, the extreme apex remaining tubular but bifid for

Fig. 30. *Roscoea ganeshensis* growing in the Ganesh Himal, Nepal. Photographed by Bill Baker, 1992.

Fig. 31. ***Roscoea ganeshensis.*** BBMS 50. Cultivated at Kew and photographed by Andrew McRobb, R.B.G., Kew (comm. Tony Hall).

2–5 mm. *Inflorescence* non-pedunculate. *Flowers* mid-purple, with or without dark purple spot at base of labellum. *Perianth tube* purple, widening at joint with perianth, pubescent, 8.5 cm long. *Dorsal petal* erect, outer surface whitish with 9 purple veins, inner surface mid-purple at apex, white, striped dark purple below, apiculate tip white, cucullate, 3–3.5 × 1–1.4 cm. *Lateral petals* joined centrally at base, mid-purple, paler at base, apex acute to obtuse, 2.7–3.1 × 0.6–0.9 cm. *Staminodes* spathulate, pale to mid-purple, the claw whitish, striped at base with dark purple on inner surface, raised fleshy vein near shorter edge, tips sometimes recurved towards dorsal petal, 1.6–2 × 0.5–0.7 cm, including 0.2–0.5 cm long claw. *Labellum* with crumpled surface and undulate edge, mid-purple, throat deep purple, spreading to blotch, if present, between the raised shoulders which are at right angles to and above the claw, emarginate, 3.6–4.5 × 2.5–3.5 cm including the 0.6–1 × 0.7–1 cm claw, base papillose, with 2 raised fleshy folds at middle of area above the claw, apex bilobed for 0.8–1.5 cm. *Stamen* white, 1.3–1.85 cm. *Filament* triangular, flat, striped with dark purple on inner surface at base, 4–5 × 2.5–5 mm, narrowing to 1.5–2 mm wide at join with appendages. *Anther* thecae 6–7.5 × 0.75–1 mm, pollen cream; connective bilobed at apex, just above thecae; appendages cream, pointed, in line with connective, 1.5–3 mm long; filament and connective area 5–8 mm long to join with thecae. *Style* 10 cm long. *Stigma* helmet shaped, apex hairy, 1 × 1 mm.

5. ROSCOEA NEPALENSIS

This species was described in 1980 from herbarium specimens of three gatherings in 1952 made by Polunin, Sykes & Williams, and five made in 1954 by Stainton, Sykes & Williams. It seems, though, that no living plants were collected for cultivation, so that no painting of the plant in flower has been possible. Oleg Polunin did take some photographs and one of these is reproduced here to illustrate this species. Half a century had to pass before this beautiful species was introduced to cultivation. An article by Edward Needham in *The Alpine Gardener* volume 70 (2002), reports that plants he has collected recently 'present no difficulties in cultivation'.

Roscoea nepalensis has only been recorded from a small area to the west of central Nepal, between Padmara and Jumla, Talphi, Ranagaon and Lete, south of Tukucha on the Kali Gandaki river. Edward Needham suggests that it may occur further to the west of these localities. A single collection made by G. Miehe in 1977 also came from near Lete. He described the plant's habitat as being mossy pine forest, similar to that described by Edward Needham. Miehe declared the flowers as being 'brightly white' and so far no coloured forms of this species have been recorded, and it would seem to be the only species in the genus that is normally white-flowered. Many other species whose flowers are usually coloured include the occasional albino within populations.

Roscoea nepalensis has no clear affinities with species within or outside its known geographical range. The anther is most like that of *R. alpina*, with very small appendages held almost immediately below the base of the thecae. The basal elongation of the connective is minimal, and not at an angle to the thecae. The elliptic dorsal petal is small and narrow, and only two thirds the length of the labellum. The linear-oblong lateral petals are the same length as the dorsal petal, whereas the obovate labellum is comparatively very large, with a V-shaped cleft for a quarter to half its length, and lacks a claw. The large staminodes are unusual in being almost circular; furthermore, their main vein is more or less central, and is not prominently raised.

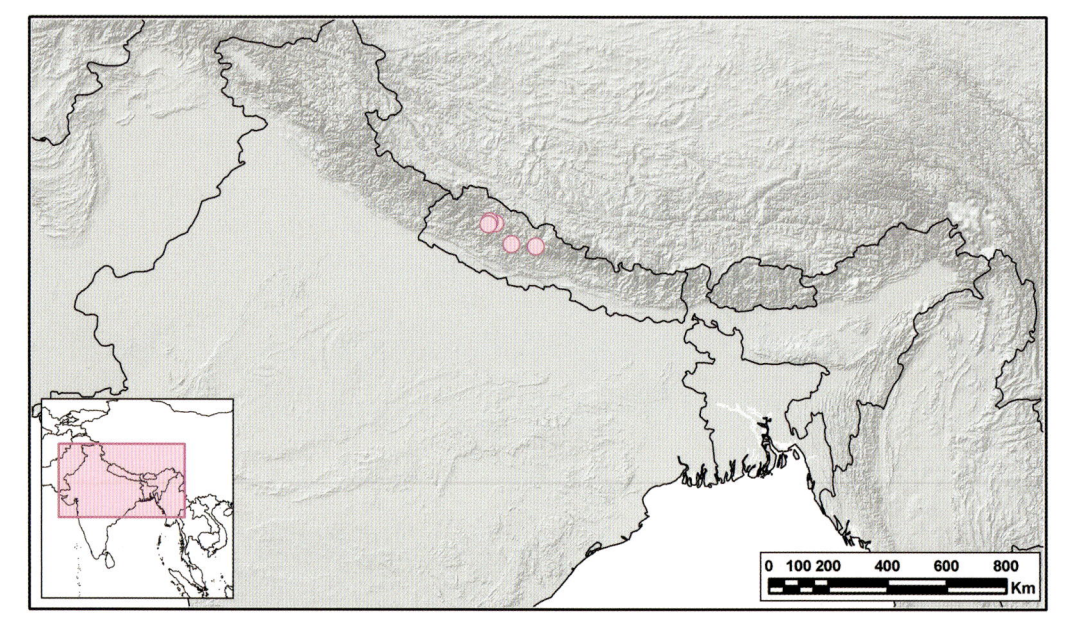

Map 8. Distribution of ***Roscoea nepalensis***.

The leaves are not distichously arranged, but are more or less rosulate. The leaf bases are narrow and petiole-like, as in the Chinese species, *Roscoea debilis*.

This species grows in open grassy and rocky habitats and in shady situations in open mossy pine forest. At present it has only been recorded from northern central Nepal in the States of Jumla, Piuthan and Palpa, and in the Kali Gandaki river area, at altitudes between 2300 and 3050 m; it flowers in June and July.

Roscoea nepalensis Cowley in Kew Bull. 34(4): 811 (1980). Type: Nepal, N of Jumla, 2440–2740 m, 29 June 1952, *Polunin, Sykes & Williams* 4391 (holotype BM; isotypes E, K).

R. alpina sensu Hara in Hara *et al.*, Enum. Fl. Pl. Nepal 1: 61 (1978) *quoad Polunin, Sykes & Williams* 362.

DESCRIPTION. *Perennial herb* 10–26 cm tall. *Sheathing leaves* 1–3, soon splitting. *Leaves* 2–5, ovate to ovate-lanceolate, (5–)7–10(–15) × (0.8–)1.4–2.6(–3.4) cm, acute, with a short, narrow, petiole-like base, veins diverging from the pale keel, ligule area ± semicircular. *Inflorescence* exserted, non-pedunculate, flowers white. *Bracts* ± equal to the calyx, 4–8 × 0.6–0.8 cm, acute, pale green. *Calyx* bluntly bidentate. *Perianth tube* 6–9 cm long, exserted from the calyx. *Dorsal petal* elliptic, 2.2–3 × 1 cm, cucullate, prominently veined. *Lateral petals* linear-oblong, 2–2.5 × 0.5 cm, obtuse. *Labellum* very broadly obovate to obovate, 3.5–4 × 2.5–4 cm, not deflexed, not clawed, bilobed, each lobe emarginate, edges undulate. *Staminodes* ± circular, 1.5–2.3 × 1.2–1.5 cm, vein ± median, weak; claw short. *Anther* cream, thecae 8 × 3 mm, basal connective elongation very short, appendages 1–2 mm. *Stigma* erect, galeate. *Epigynous glands*, *capsule* and *seeds* not seen.

Fig. 32. **Roscoea nepalensis.** Photographed in the type locality by Oleg Polunin.

68 THE GENUS ROSCOEA
PLATE 6

6. ROSCOEA TUMJENSIS

When I wrote about *Roscoea ganeshensis* in the *Botanical Magazine* in 1996, I said that "the little-known species *R. tumjensis* Cowley also occurs in the Ganesh Himal, but has not been recorded since the type collection." At the time this was correct, because the plant illustrated here, had not yet flowered and shown its true identity, although it had been collected at the same time as *R. ganeshensis*. When it was collected by the Oxford University Ganesh Himal expedition team (Baker, Burkitt, Miller & Shrestha) in 1992, it was tentatively identified as *R. purpurea*. Later, when introduced in cultivation at Kew, its prominently auriculate leaf bases suggested that it might be *R. auriculata*, even though this would have implied an extension to the known distribution of that species. This collection (BBMS 15) kept everyone in suspense for three years before it flowered; it was then positively identified as *R. tumjensis*.

When the team members saw this species in the wild, they described the flowers as occurring in "an excellent array of colour forms, most notably, an intense inky purple swarm". According to the data accompanying the plants which were brought into cultivation, they had "large, very dark purple flowers" and the habitat was described as being a "dangerously precipitous, open grassy slope on the edge of mixed (mostly Rhododendron) woodland".

As luck would have it, the first flowering at Kew in June 1995, was the best display it has ever produced; one perfect flower was open simultaneously on each of the three plants in a single pot. Unfortunately Kew was hosting an international conference at the time, and neither a photographer nor artist could be found to capture this moment of triumph, so the chance was lost for that year. Two years later the plants flowered again and fortunately Christabel King was available to paint this treasured plant at the critical moment, even though we had to be content with a single flower.

I have seen only a few herbarium specimens of wild plants, but these show that this species is similar in habit and general flower form to the commonly cultivated Chinese species *Roscoea humeana*. Both start to flower before the leaves are fully developed. However, this is seldom the case in *R. humeana* or in *R. tumjensis* in cultivation, as can be seen from the illustrations (Figs. 35 & 78). The herbarium specimens were collected by P.C. Gardner from around Tumje in 1953 while he was a member of the New Zealand Himalayan Expedition. They show robust plants with large, concolorous, purple flowers with wide perianth segments and stems bearing short under-developed leaf blades with well defined auricles at their bases. He made two collections, both housed in the Natural History Museum in London: *Gardner* 525 was collected in May at 2900 m, and *Gardner* 790, collected in June at 2740 m. The latter specimen was chosen as the type and there is a duplicate of the collection in the herbarium of the Royal Botanic Garden, Edinburgh. *McCosh* 65, collected at Junbesi in May 1964 at 3050 m, comprises handsome specimens with large deep purple flowers, is also at the Natural History Museum, as is *Stainton* 4594 from Solu Khola in East Nepal, collected in June 1964 at c. 3200 m. Stainton reported seeing the plants growing singly or in groups of up to twenty, amongst shrubs. The earliest collection was probably made by *Capt. Lall Dhwoj* (no. 04) in 1930 at Shatey between 2500 and 3600 m. However, in the Herbarium of the Honourable East India Company (the Wallich Herbarium) at Kew, on a sheet of *Wallich* 6528A that bears several specimens, mostly *R. purpurea*, there is a single plant that is probably this species. This would have been collected at the beginning of the 19th Century, between 1815 and 1822, and probably during the latter year.

Plate 6. *Roscoea tumjensis*. BBMS 15. Painted by Christabel King, July 1997.

6. ROSCOEA TUMJENSIS

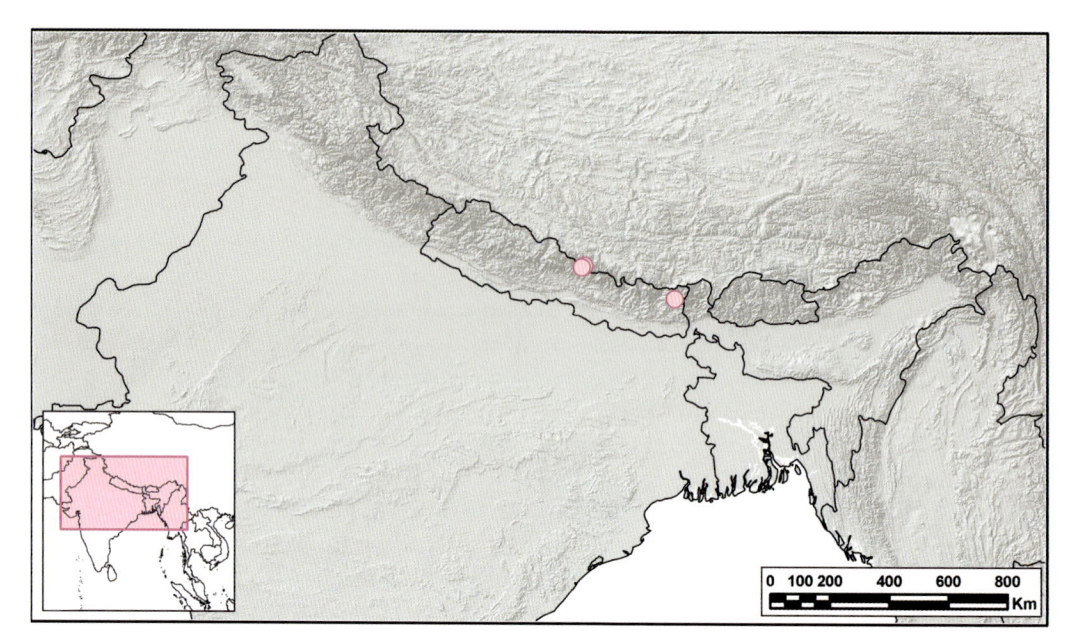

Map 9. Distribution of *Roscoea tumjensis*.

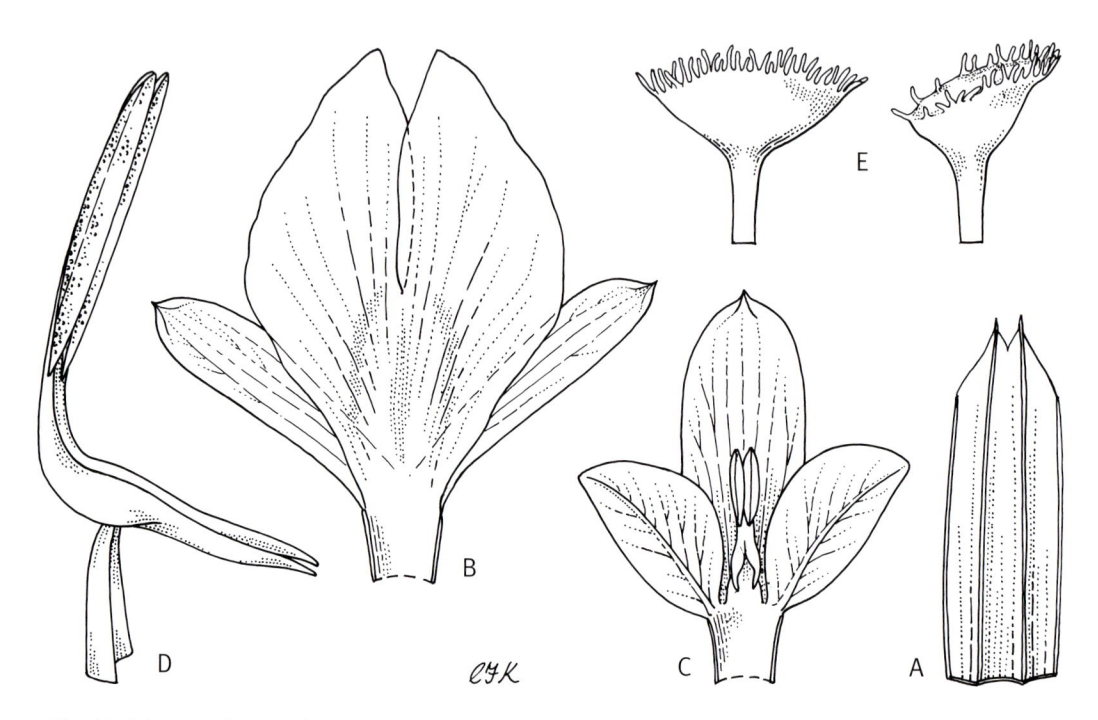

Fig. 33. Dissection diagram of *Roscoea tumjensis* by Christabel King. **A** upper portion of calyx opened out; **B** lower portion of perianth; **C** upper portion of perianth and stamens; **D** stamen; **E** stigmatic surface, abaxial and side view.

Hara obviously had trouble with these collections when he was preparing "*An Enumeration of the Flowering Plants of Nepal*" (Hara et al., 1978). He cited *Gardner* 525 as *Roscoea purpurea* var. *auriculata*, and also as *R. chamaeleon*. He also assigned *McCosh* 65 to the latter species, but annotated the specimen "cf. *R. humeana*". My later studies confirmed that these collections constituted an undescribed species.

The locality from which the collection of *Roscoea tumjensis* illustrated here originated is obviously rich in species of *Roscoea*; two other species were found in the same place on the Ganesh Himal Expedition: *R. alpina* (BBMS 14) and *R. capitata* (BBMS 13). In all, five out of the seven known Nepalese species were found in the Ganesh Himal by this team. The other two grow further west, in the Kali Gandaki area (*R. nepalensis*) or further east, on the Sikkim border (*R. auriculata*).

A character that shows clearly in cultivated plants but which is difficult to see in herbarium collections, is the long corolla tube which when exserted makes the flower nod, as can be seen in this illustration. However, Edward Needham states (*Bulletin of the Alpine Garden Society* 2002) that this trend is not evident in all the plants of *Roscoea tumjensis* that he has in cultivation. His plants came from eastern Nepal, and he reports that they are the earliest of the Himalayan species to flower and become the tallest and most robust of the roscoeas, eventually becoming over a metre tall.

In the second edition of *Flowers of the Himalaya* (Polunin & Stainton, 1997), photograph 1379 on Plate 119 is labelled '*Roscoea purpurea* from Tamur, East Nepal'. The plant depicted is certainly not *R. purpurea*, and does not show the exserted bracts of *R. auriculata*, which occurs in the extreme east of Nepal and throughout Sikkim. It does, however, show the auriculate bases of the leaves and the concolorous perianth segments, both characteristics of *R. tumjensis*. It is difficult to positively identify *Roscoea* species from photographs but it is more than likely that this photograph is of this species.

Fig. 34 (left). **Roscoea tumjensis**. Photographed in the Ganesh Himal, Nepal by Bill Baker.
Fig. 35 (right). **Roscoea tumjensis**. Cultivated in the Alpine House at Kew and photographed by Richard Wilford, R.B.G., Kew.

Table to show comparisons between three related species of *Roscoea* found in Nepal

	R. auriculata	*R. purpurea*	*R. tumjensis*
Leaf and base	In the wild state, leaves well developed at flowering time; all leaf bases clearly auriculate.	In the wild state, leaves well developed at flowering time; most leaf bases not auriculate, but sometimes the lower leaves slightly auriculate.	In the wild state, leaves not fully developed at flowering time; all bases clearly auriculate.
Epigynous glands	4–5.5 mm long	5.5–9.5 mm long	[No data].
Bracts	Exserted above leaves, equal to or slightly shorter than calyx.	Apex of bracts only exserted above leaves, longer than calyx.	Not exserted above leaves, much shorter than calyx.
Staminodes	Usually white, 1.5–2 × 0.6–0.9 cm, asymmetrically obovate or rhombic with short claw; vein excentric.	White or mauve veined white, 2.5–4 × 0.6–1.1 cm, obliquely spathulate, narrow with a long claw; vein excentric.	Concolorous, 1.1–2.2 × 0.8–1.3 cm, circular to elliptic with very short claw; vein ± central.
Anthers	Connective elongation in line with rounded appendages; connective & appendages 4–6 mm long.	Connective elongation angled to pointed appendages; connective & appendages 9–25 mm long.	Connective elongation in line with rounded appendages; connective & appendages 6–10 mm long.

The characters that distinguish this species from the very similar species *Roscoea purpurea* and *R. auriculata*, are: leaf development and shape of leaf base, size of epigynous glands, habit of the bracts, shape and size of staminodes and anthers. These characters are summarised in the table above.

At present this species is known only from central Nepal, in an area to the west of Kathmandu, and further east around the Dudh Kosi valley area, Tamur and Solu Khola. Plants have been found growing in grassy areas on rocky hillsides, and in clearings in woodland or forest at altitudes between 2500 and 3600 m, and flowering from April to July.

Roscoea tumjensis Cowley in Kew Bull. 36 (4): 755–756 (1982). Type: Nepal, Shiar Khola river, above Tumje, 2740 m, 15 June 1953, *Gardner* 790 (holotype BM, isotype E).

DESCRIPTION. *Rhizomatous, tuberous-rooted herb* to 50 cm. *Sheathing leaves* 1–4, obtuse to truncate, sometimes strongly marked purple-red. *Leaves* up to 7, oblong to ovate, 4–32 × 1.5–4.5 cm, obtuse, widest at the auriculate base, mid-green and punctate above, paler below, scarcely developed at flowering time or flowering precociously, except in cultivation; side veins parallel. *Stem* formed by overlapping reticulate leaf sheaths, elliptic in section. *Ligule* raised for 1–2 mm, ligule area ± semicircular. *Inflorescence* not exserted, flowers pale lilac, lilac-blue, pinky purple, bright purple to very dark purple. *Bracts* much shorter than the calyx, acute, ciliate. *Calyx* tubular below, bilobed or tridentate, lobes broad, rounded, ciliate, apiculate, apices keeled, 1.5 cm. *Corolla tube* exserted to well exserted up to 3 cm from the calyx at maturity, white below, graduating to coloured beneath the flower. *Dorsal petal* obovate to broadly elliptic, 2.7–3.5 × 1.2–2.7 cm, apiculate. *Lateral petals* oblong, 2.6–3.5 × 0.5–1.4 cm. *Labellum* very broadly obovate, 3.2–5.5 × 2.2–4.5 cm including a short claw to 7 mm, edges undulate, emarginate, bilobed for 1.3–3.2 cm. *Staminodes* circular to elliptic, 1.1–2.2 × 0.8–1.3 cm, concolorous, with the vein ± in the middle, very shortly clawed to 3 mm. *Filament* 4–5 × 2.5–3 mm. *Anther* cream, angled below the thecae, thecae 6–10 × 1.5–3 mm, connective elongation 3–5 mm, in line with the thin, curved, obtuse appendages ± 5 mm long. Fruit and seeds not seen.

7. ROSCOEA AURICULATA

Roscoea auriculata is probably the easiest of the species to establish in cultivation, but for many years it has been grown in gardens under the name '*R. purpurea*'. Of course, there is always a temptation to apply the name 'purpurea' to any purple *Roscoea* that may come into cultivation There are, however, quite clear differences between the two Himalayan species that were recognised by Karl Schumann when he described this species in 1904 (see comparisons in the previous species account of *R. tumjensis* (p. 72)). This species had been collected many times before 1904 from areas around Mount Everest and eastern Nepal, but the majority of the collections came from Sikkim.

Schumann used an un-numbered specimen from Sikkim, supposedly collected by Joseph D. Hooker and Thomas Thomson between 2250 and 3000 m. However, Hooker was alone or accompanied by the British Resident, Dr Campbell, when in Sikkim. Thomson only joined him after his release from captivity in Sikkim and the two men then travelled together in Assam. However, the specimen cannot be checked; it was presumably lost when the bulk of the Berlin Herbarium was destroyed in World War II, and no duplicate with identical annotations has found. In 1982 I chose a Hooker collection from Lachen in Sikkim as a neotype; this was collected in 1849, and is held in the herbarium at Kew.

Roscoea auriculata was introduced into cultivation by Henry John Elwes, who first went to Sikkim in 1870 but did not introduce this plant until 1881. Elwes was obviously aware that his plant was different from J.E. Smith's original *R. purpurea*, as he wrote in the *Gardener's Chronicle* in 1890 "…..I have distributed under the name of *R. sikkimensis* Hort. Elwes, as C. B. Clarke told me it had no name other than *R. purpurea*……". He goes on to discuss the differences between his introduction and the original, including the information that in Sikkim it can grow as an epiphyte and as a result has a different root structure to that of *R. purpurea*. He said that he sent seeds to Mr Thompson of Ipswich who raised it and supplied plants. The plants now found commonly in gardens may originate from this source. On p. 191 of the same volume, where Elwes' letter to the editor is published, Fig. 30 is labelled *R. purpurea*, but depicts *R. auriculata*. On page 251, in reply to Elwes' letter, W. Thompson of Ipswich confirms Elwes' previous statement about the differences between the species and later says "The colour too, is strikingly different, (*purpurea*) being a very pale lilac; by no stretch of meaning can it legitimately be called purple. The specific name is far more appropriate to the Sikkim variety, which bears flowers of the deepest violet-purple, somewhat smaller and with slenderer tube…..". It must be remembered, however, that J.E. Smith did not have any other species for comparison when he described *R. purpurea*; it was the first species described in the new genus named after William Roscoe of Liverpool.

At Kew there is an herbarium specimen, labelled *Roscoea purpurea*, of the Elwes plant that was illustrated (Fig. 30) in the 1890 *Gardener's Chronicle*, as well as one of cultivated *R. auriculata*, dated

74 | THE GENUS ROSCOEA
PLATE 7

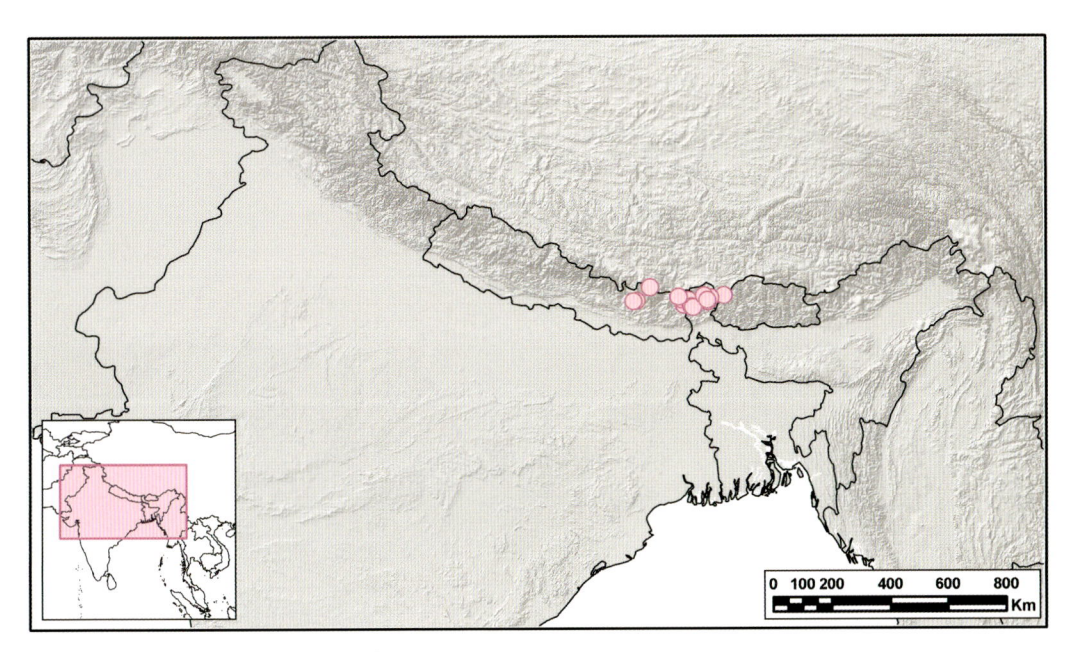

Map 10. Distribution of **Roscoea auriculata**.

29 July 1919 (Kew accession number H.879-19) which was shown at the Royal Horticultural Society meeting by H.J. Elwes. For some reason this specimen is labelled 'Roscoea purpurea capitata'. Both of these specimens are also labelled as coming from Darjeeling.

Confusion continued over these two species for some while; people were growing *Roscoea auriculata*, but calling it *R. purpurea* for want of a legitimate name. In 1900 S. Arnott remarked in the *Gardener's Chronicle* that he had seen it growing in several gardens "notably in the Daisy Hill Nursery at Newry, the Glasnevin Botanic Garden in Dublin and at Mount Usher in Co. Wicklow".

In 1907, Louis Gentil published *Roscoea sikkimensis* in his *Liste des Plantes cultivées dans les Serres chaudes et coloniales du Jardin botanique de l'État à Bruxelles*, thus validating Elwes's epithet. However, he was too late: Schumann had published the name *Roscoea auriculata* in 1904, and this, being the earlier, is the name that must be used. In spite of this, Herbert Maxwell in 1928, and many authors writing in the *Gardener's Chronicle* until the 1970s, continued to refer to this plant as *R. purpurea*.

In 1978, in the *Enumeration of the Flowering Plants of Nepal*, Schumann's species was relegated to a variety of *Roscoea purpurea*, but I consider that there are sufficient differences in the size and/or shape of the leaves, bracts, flowers, staminodes, anther, and epigynous glands to justify maintaining *R. auriculata* as a good species. In the 1996 *Flora of Sikkim*, both *R. purpurea* and *R. auriculata* are cited as coming from the same areas of Lachung, Bakhim and Lamteng; the latter species is listed as growing in Chungthang and Panthgang.

In 1983 the Alpine Garden Society organised a successful seed collecting expedition to Sikkim even though the area, especially northern Sikkim, was closed to foreigners. They found *Roscoea auriculata* in Yoksum, and there is a good photograph of it on page 268 in the special expedition

Plate 7. *Roscoea auriculata* with habit sketch. Cultivated at R.B.G., Kew. Painted by Christabel King.

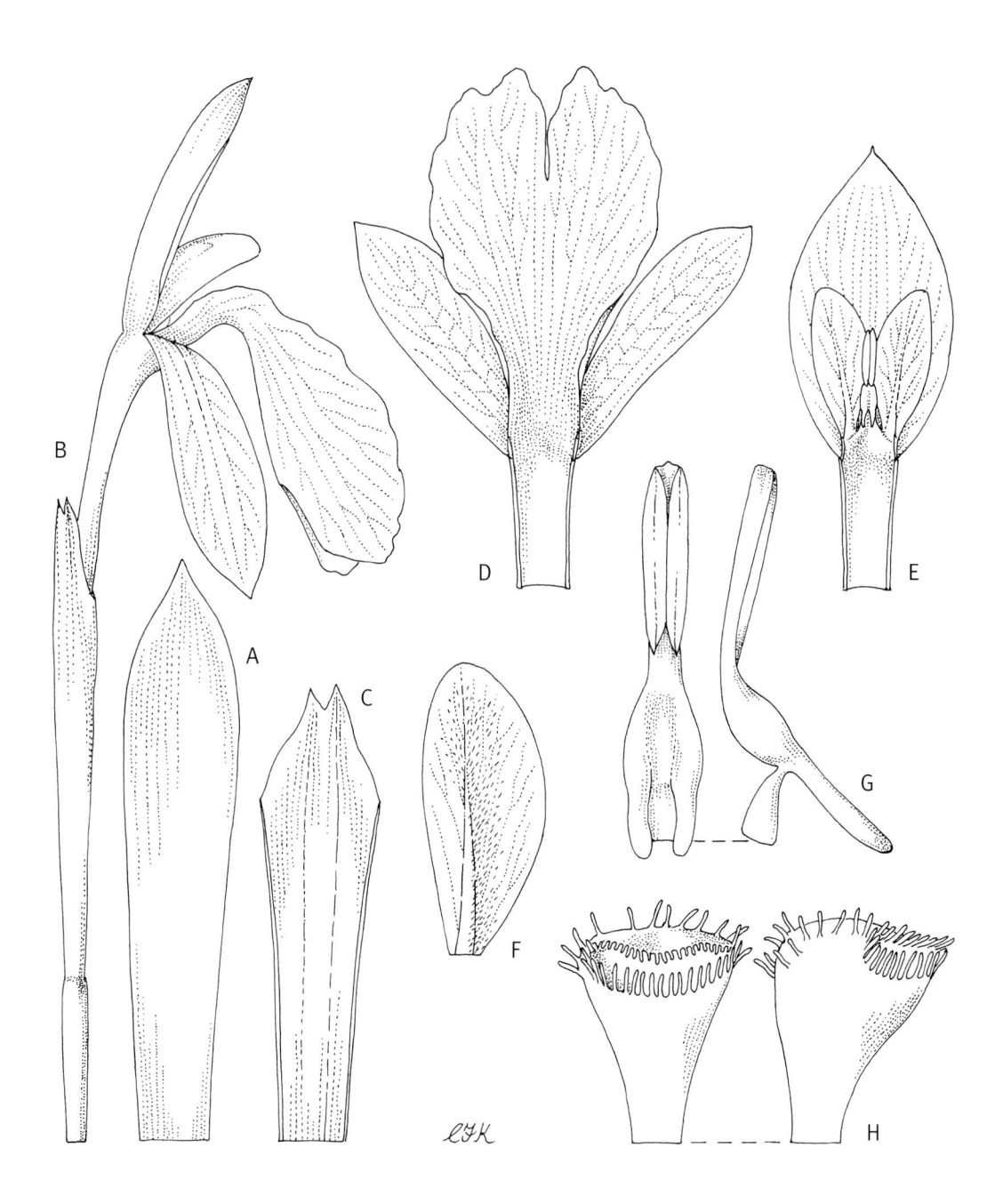

Fig. 36. Dissection diagram of ***Roscoea auriculata*** by Christabel King. **A** outer bract; **B** flower; **C** inner bract; **D** lower part of corolla; **E** upper part of corolla; **F** inner perianth segment; **G** stamens; **H** stigma, 2 views.

THE GENUS ROSCOEA
7. ROSCOEA AURICULATA

Fig. 37. Habitat of *Roscoea auriculata* growing between Bakim and Yoksum, near the Prechu river, Sikkim, c. 2800 m on 5 July 1983. AGSES 319. Photographed by Brian Mathew.

Fig. 38. *Roscoea auriculata* growing between Bakim and Yoksum, near the Prechu river, Sikkim, c. 2800 m on 5 July 1983. AGSES 319. Photographed by Brian Mathew.

report edition of the Alpine Garden Society's Journal in September 1984 (reproduced here, p. 77). In 1987 David Lang followed in that expedition's footsteps and visited the Lhonak valley. In the same journal he reported seeing *R. auriculata* at Lachen, a favourite collecting area of Joseph Hooker's in 1849. Lang commented on the "luscious" *R. auriculata* in that area.

Roscoea auriculata is certainly an extremely handsome plant, especially when a mature clump of plants is in full flower, displaying the exposed bracts and the fine white staminodes, set against the deep purple flowers. Various forms of this species are now sold by nurseries, namely, 'White Cap', 'Floriade' and 'auriculata-pink form'.

Roscoea auriculata can be found in eastern Nepal and Sikkim and in the areas of southern Tibet to the north of those countries. Plants grow on roadside banks, rocky ground and in forest clearings from 2130 to 4880 m, and flower between May and September.

Roscoea auriculata K. Schum. in Engl., Pflanzenriech. 4 (46): 118 (1904). Type: Sikkim, 2250–3000 m, *Hook. f. & Thomson* s.n. (holotype B, destroyed). Sikkim: Lachen, 3050 m (10,000 ft.), 1 June 1849, *Hooker* s.n. (neotype K).

R. sikkimensis Hort. ex Gentil, Pl. Cult. Serres Jard. Bot. Brux.: 169 (1907), nom. nud.; F.W. Harvey (Ed.), Garden 78: 159, t. 158, fig. 1 (1914).

R. purpurea Sm. var. *auriculata* (K. Schum.) Hara, Enum. Fl. Pl. Nepal 1: 61 (1978), excl. *Kanai, Hara & Ohba* 723607 [as TI 723601].

Fig. 39 (left). **Roscoea auriculata.** Cultivated and photographed by Roland and Gay Bream, Cruckmeole.
Fig. 40 (right). **Roscoea auriculata**. Cultivated at Kew. Photographed by Richard Wilford, R.B.G., Kew.

ILLUSTRATION. *The Rock Garden* 26(4): fig. 113; text p. 322 (2000).
DESCRIPTION. *Plants* (20–)25–42(–56) cm tall. *Sheathing leaves* 1–2. *Leaves* (3–)5–7(–10); lower leaves obovate, 5–27 × 1.5–6 cm, usually widest at the middle, linear to broadly elliptic, acute to acuminate, auriculate at base, mid yellow-green above, grey-green beneath, glabrous, closely veined with side veins ± parallel. *Leaf sheaths* reticulately veined, sometimes flushed red towards base of stem. *Ligule* area ± semicircular, ligule raised for 0.75 mm. *Internodes* (1–)3–10 cm long. *Inflorescence* with no exserted peduncle. *Flowers* deep purple or occasionally white, or with some parts white. *Bracts* equal to or slightly shorter than the calyx, (6–)7–8(–10) × 0.6–2 cm, acute, first bract positioned below main inflorescence, sometimes second and third bracts fused together, pale green, sometimes tinged with brownish-pink streaks. *Calyx* 5–8 cm long, sharply or bluntly bidentate, ciliate, brown markings concentrated more towards apex. *Corolla tube* 6.5–8 cm long, ± same length as calyx, but occasionally well-exserted. *Dorsal petal* obovate to broadly elliptic, (2.8–)3–3.5(–3.7) × (1.5–)1.9–2.5(–2.8) cm, cucullate, apiculate. *Lateral petals* elliptic to linear-oblong, (2.3–)2.5–3.5 (–3.8) × (0.6–)0.8–1(–1.2) cm, acute, sometimes fused together. *Labellum* very broadly obovate, 3.3–4.8 × 2.5–4 cm including 0.7–1.2 cm claw, deflexed, entire to lobed for half length of limb, 6–8 faint white lines running down claw into throat. *Staminodes*, asymmetrically obovate, 1.5–2 × 0.6–0.9 cm, usually white with V-shaped purple markings at base, vein excentric. *Anther* white, angled below the thecae, thecae 7–9 × 1.5–3.5 mm, connective elongation in line with the pointed yellow appendages, joint of connective and appendages swollen, together 6–8 mm long. *Pollen* cream. *Ovary* 1–2.5 × 0.2–0.4 cm. *Epigynous* glands 4–5 mm long. *Capsule* 2–3 × 0.7–0.8 cm. *Seeds* brown, slightly constricted near middle, aril shallowly lacerate.

8. ROSCOEA BHUTANICA

This species is the most recently described member of the genus, and the latest to be introduced into cultivation. It is reminiscent of a diminutive *Roscoea purpurea*.

When I published my revision of the genus in 1982, I had seen little material from Bhutan, either pressed or living; at that time the country was virtually closed to foreigners. At the time I felt that the little material that there was of this taxon seemed to be best placed in the Chinese species *Roscoea tibetica*. However, in my account of *R. tibetica* I pointed out that there were differences between the Chinese and the Bhutanese specimens, and warned that my conclusions might have to be revised if and when more material became available. Two other species had been recorded from Bhutan, *R. alpina* and *R. purpurea*. The form of *R. purpurea* found in Bhutan has larger flowers than the one found in the middle Himalayas, and is closer in size to the forms of the same species found in the west of its range. The forms at the extremes of the range also have very pale flowers; the one from Bhutan is white with dark purple streaks on the labellum. This form is in cultivation, introduced by Kingdon Ward from Assam, and now known as *R. purpurea* forma *alba*. In the past it was often wrongly called *R. procera*.

The *Flora of Bhutan* is an ongoing project compiled by scientists at Edinburgh Botanic Garden. For many years the botanists were unable to enter the country to study the plants *in situ* that they were writing about. Happily, these restrictions have now been lifted and as a consequence, further, more detailed field studies have been possible. During a PhD project that included a detailed phylogenetic study of the genus *Roscoea*, it became clear that the Chinese plants of *R. tibetica*, not

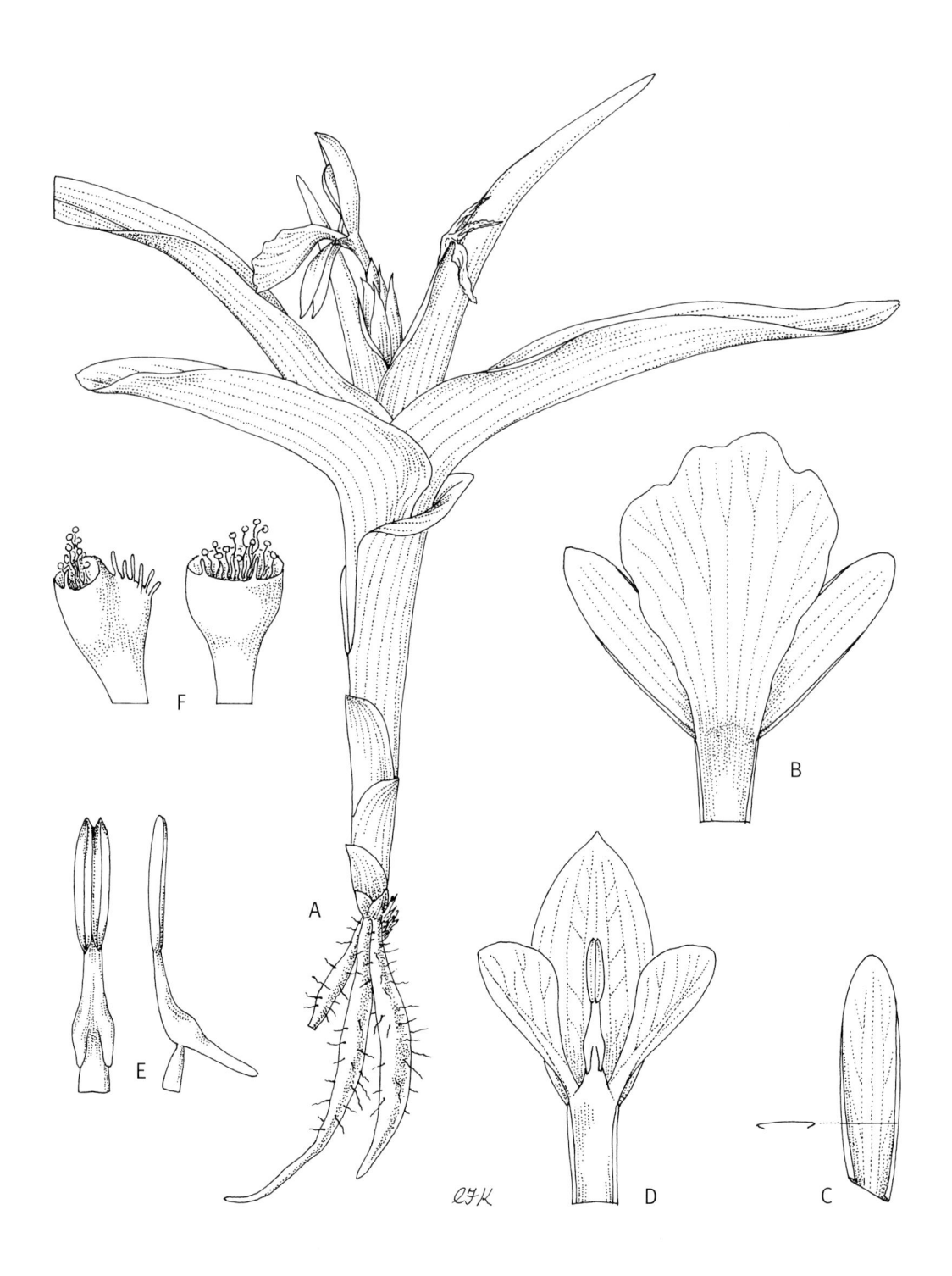

Fig. 41. Dissection diagram of **Roscoea bhutanica** by Christabel King. **A** habit; **B** lower part of corolla; **C** lateral petal; **D** upper part of corolla; **E** stamen with germinating pollen grains; **F** stigma, 2 views.

8. ROSCOEA BHUTANICA

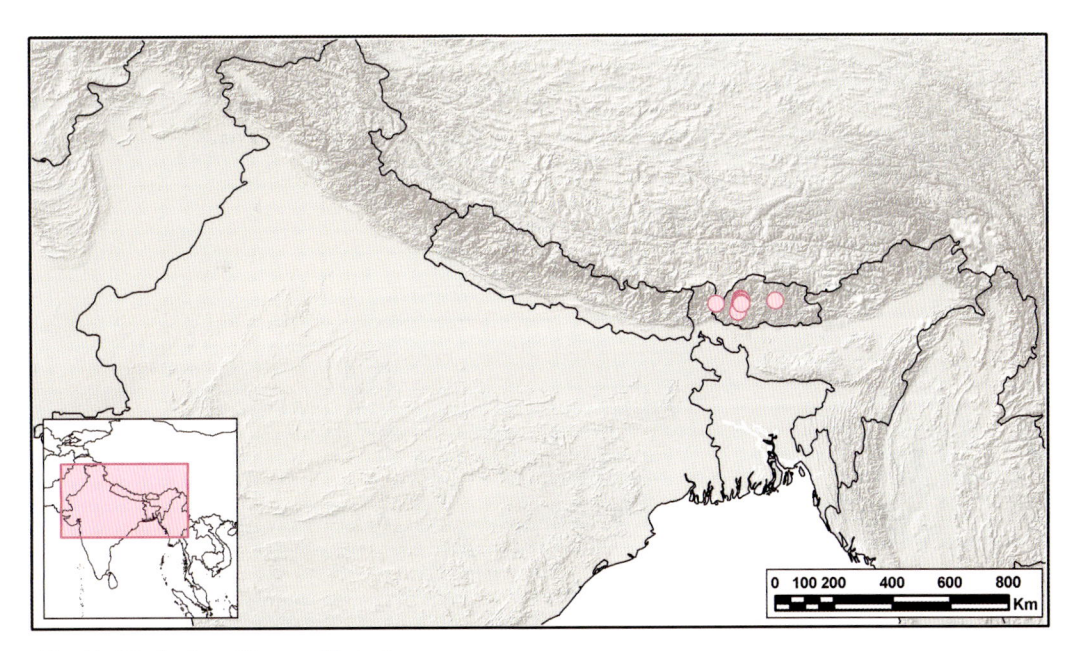

Map 11. Distribution of ***Roscoea bhutanica***.

Fig. 42. ***Roscoea bhutanica*** growing near Dochu La, Bhutan, 2nd July 2002. Photographed by Bill Baker.

8. ROSCOEA BHUTANICA

only had a disjunct distribution, but also had characters that did not match those of the Bhutanese plants. Further studies revealed that they represent a species that had not previously been described. C. Ngamriabsakul and M.F. Newman described *R. bhutanica* in 2000 using plant material collected in the Bumthang district of Bhutan in 1979 by Grierson & Long, botanists from Edinburgh.

Molecular systematic studies place this new species in the Himalayan clade, close to *Roscoea purpurea* and *R. auriculata*. Dissections of *R. bhutanica* show that the shapes of the floral parts are proportionately very similar to *R. purpurea*, with the clawed staminodes nearly as long as the dorsal petal, and a long connection between the anther thecae and the appendages. The whole plant, however, is smaller and somewhat squat; the leaves are crowded together, hiding the stem, not well spaced out as is typical in *R. purpurea*.

Ngamriabsakul & Newman's paper gives morphological comparisons between *Roscoea bhutanica* and *R. tibetica*. The main differences are as follows. In *R. tibetica* the corolla tube is long relative to the calyx and exserted from it, while in *R. bhutanica* the corolla tube is short and entirely enclosed within the calyx. In *R. tibetica* the calyx is longer than the bract while in *R. bhutanica* it is shorter than or equal to the bract. The labellum is shorter than the lateral petals in *R. tibetica*, but longer than them in *R. bhutanica*. *R. tibetica* can flower precociously whereas *R. bhutanica* does not normally flower until several leaves have been produced. In *R. tibetica* the leaves are arranged in a rosette and in *R. bhutanica* they are distichous.

Roscoea bhutanica has been recorded found in five out of the twenty-three botanical districts of Bhutan, as defined in the *Flora of Bhutan*, and in southern Tibet to the north of Bhutan. It grows in pine forests, wooded valleys, on open, sunny grassy banks, in meadows and in sheltered clearings in forests. Plants are to be found at elevations between 2290 and 3600 m, and flower between May and August.

Fig. 43. ***Roscoea bhutanica*** growing at NE Wangdi, Bhutan. Photographed by Phil Cribb, R.B.G., Kew.

Roscoea bhutanica Ngamriab. in Edinburgh J. Bot. 57 (2): 271–278 (2000). Type: Bhutan, Bumthang Distr., Bumtang Chu, Byakar, wooded valley above Lami Gompa, 3050 m, 12 June 1979, *Grierson & Long* 1826 (holotype E).

R. tibetica sensu Cowley (1982), quoad spec. *Cooper* 1300, *Gould* 912, *Grierson & Long* 116, *King's Collector* 454, *Ludlow & Sherriff* 2275, *Ludlow, Sherriff & Hicks* 16377.

DESCRIPTION. *Tuberous rooted herbs* 7–14(–22) cm tall. *Sheathing leaves* 2–4, apex obtuse. *Leaf blades* usually 2–6 at flowering time, lanceolate-ovate to oblong, 4–21(–30) × 1–5 cm, slightly auriculate, glabrous, crowded together at the base. *Inflorescence* enclosed in leaf sheaths. *Flowers* opening just above leaves, all shades of purple, one open at a time. *Bracts* oblong to spathulate, 4.5–8 × 1–1.6 cm, acute. *Calyx* more or less equal to bract, split by 1–1.5 cm, apex bidentate, teeth 1–3(–9) mm long. *Corolla tube* usually longer than calyx by up to 1 cm, rarely equal to or shorter than it, 5–6.5 cm long. *Dorsal petal* narrowly oblanceolate, 2–3 × 1.1–1.3 cm, apiculate. *Lateral petals* linear-oblong, 2.4–2.8 × 0.4–0.6 cm, obtuse. *Labellum* slightly deflexed, obovate, 2.5–3.2 × 1.6–2 cm, entire or lobed for more or less half its length, without white lines at claw. *Staminodes* obliquely spathulate, 1.6–1.9 × 0.5–0.6 cm. *Anther* white, thecae 6–7 mm long, at right angles to connective elongation and pointed appendages. *Ovary* 1–1.7 × 0.3 cm. *Epigynous glands* 4–5 mm. *Style* pinkish-white. *Stigma* white. *Seed aril* shallowly lacerate.

9. ROSCOEA BRANDISII

The illustration of this species (Plate 8) is reproduced from the *Botanical Magazine* (t. 4630), published in 1852. Over 150 years have passed since Sir Joseph Dalton Hooker introduced this plant to cultivation from Assam, but sadly it has long since been lost. Hooker sent tubers of *Roscoea brandisii* from Khasya (Khasia) to Kew from his nine-month-long expedition to Assam in 1850, when he was accompanied by Thomas Thomson. He commented that the plant was 'very like the Sikkim one' (*R. auriculata*); his previous expedition had been to that area. However, the only name available to him at this time was *R. purpurea*, and it was under this name that the article and illustration were published.

Assam was a popular country among plant hunters in the first half of the nineteenth century. Before Hooker's expedition, William Griffith had found *Roscoea brandisii* in the Khasia Hills (*Griffith* 5736) and in 1872, C.B. Clarke found this species at Nonkrem (*Clarke* 17590A), and again in the Shillong peak area in 1886 (*Clarke* 44607A). The species was collected again in 1949 by Kingdon Ward in the Shillong peak area (*KW* 18682), in 1953 by Koelz at Mawphlang (*Koelz* 33255) and in 1973 by Tessier-Yandell, again in the Shillong area (*T-Y* 280). The former State of Assam has now been subdivided, and all the localities are in the present-day State of Meghalaya. As far as I am aware, there are no recent collections of this species. Reports that the plant had been re-introduced at the end of the 1990s were unfortunately wrong; those plants, which were from an unknown source and locality, were originally identified as *R. brandisii* but turned out to be rather weedy specimens of *R. tumjensis*, a species found only in Nepal.

As Hooker observed, *Roscoea brandisii* obviously has strong affinities with *R. auriculata*, their distributions being separated by the lowlands of northeast Assam, the Brahmaputra river and northern Bangladesh. *R. brandisii* is probably restricted to the isolated highlands of the Khasia Hills; the majority of collections come from the Shillong area. The species can easily be recognised by its narrow, linear, falcate leaves, usually without auricles at the base, and by its small, compact, purple flower with a long white perianth tube, exserted well above the calyx. The plants are also usually much smaller and more slender than those of *R. auriculata*. The curved anther with a relatively long connective elongation is reminiscent of *R. purpurea* Sm.; *R. auriculata* has a short one.

When J.G. Baker prepared the account of *Scitamineae* (including *Zingiberaceae*) for Hooker's *Flora of British India*, he recognised that the plant from Assam was different from those from further north, in the Himalayas. He found a herbarium specimen that had been annotated by G. King with the epithet '*brandisii*', but this name had never been published. Baker recognised the present species as a variety of the common Indian species *R. purpurea*, and took up King's unpublished name as a varietal

THE GENUS ROSCOEA
PLATE 8

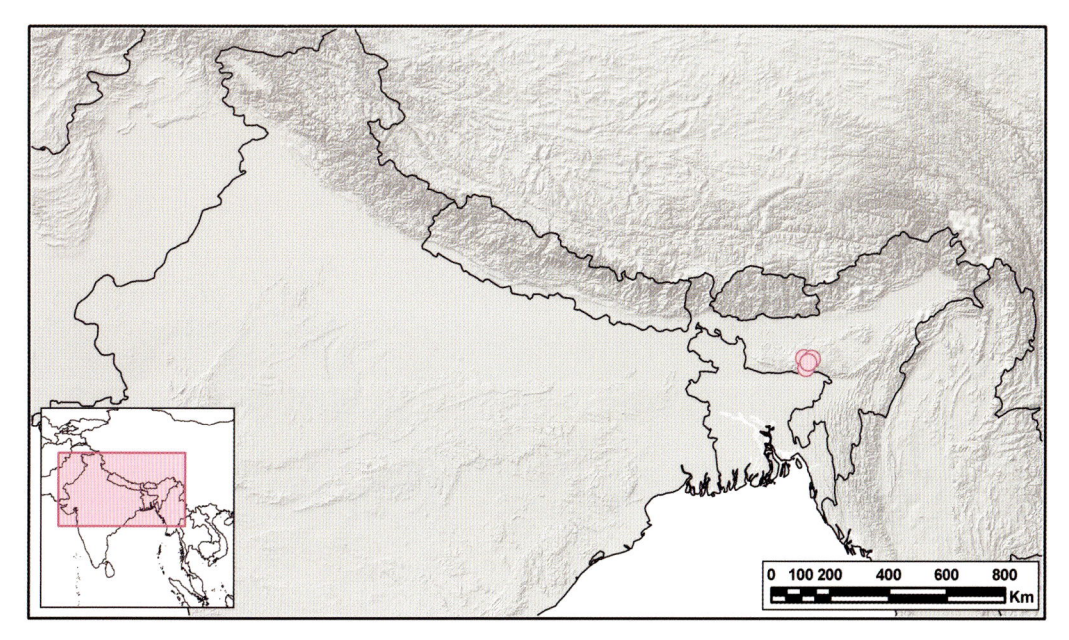

Map 12. Distribution of **Roscoea brandisii**.

epithet. K. Schumann, when drawing up the account of the genus for the fourth volume of Engler's *Pflanzenreich*, decided that the taxon deserved specific status and published it as such in 1904. The type specimen used by Baker was a *Brandis* specimen, without a number, from ?Khasia, wrongly cited as being in Burma, which had been sent to Kew from the Botanic Garden in Calcutta.

The three illustrations cited above the description (see below), show the anther to be hairy, but I have been unable to confirm this when studying herbarium specimens of this species.

This species is named after D. Brandis who was an authority on the Indian forest flora. He was of German extraction, and worked in Calcutta.

Roscoea brandisii can be found in India, the present-day State of Meghalaya, scattered in pastures and meadows, on rocky ground, on banks, hanging down cliffs and in semi-shade from 1520 to 3050 m; it flowers in July and August, the wettest time of the year when about 250 cm of rain falls in each of these months; this may go some way towards explaining why there are no recent collections!

Roscoea brandisii (Baker) K. Schum. in Engl., Pflanzenr. 4 (46): 119, fig. 16, c–e (1904) as to name only. Type as for *R. purpurea* var. *brandisii*.
R. *purpurea sensu* Hooker in Bot. Mag. t. 4630 (1852).
R. *purpurea* Sm. var. *brandisii* King ex Baker in Hook. f., Fl. Brit. Ind. 6: 208 (1890). Type: Burma, ?Khasia, *Brandis* s.n., ex Herb. Hort. Bot. Calc. 181 (holotype K, isotype CAL).
R. *exilis sensu* K. Schum. in Engl., Pflanzenr. 4 (46): 119 (1904) partly, as to description and the *Hooker* and *Clarke* specimens.

Plate 8. *Roscoea brandisii* (as *Roscoea purpurea*). *Botanical Magazine* plate 4630, September 1851, Reeve and Nichols. Painted by Walter Hood Fitch.

ILLUSTRATION. Lemaire, *Jardin Fleuriste* 3, t. 230 (1852) (as *R. purpurea*).

DESCRIPTION. *Plants* (17–)25–35(–45) cm tall, internodes long. *Sheathing leaves* 1–2, obtuse to truncate. *Leaves* 5–8(–11), linear to narrowly lanceolate, (7–)8–12(–18) × (0.6–)1–2(–2.5) cm, usually falcate, acute to acuminate, base sometimes slightly auriculate, glabrous, side veins parallel to keel. *Ligule* area ± semicircular, ligule raised for 1 mm. *Inflorescence* with short peduncle, not exserted, flowers purple. *Bracts* ± equal to or sometimes slightly longer than the calyx, acute, to 8 × 0.8 cm at fruiting stage. *Calyx* bluntly bidentate. *Perianth tube* white, long-exserted from calyx at maturity. *Dorsal petal* elliptic to broadly elliptic, 2.3–2.8 × c. 1.3 cm, cucullate, apiculate. *Lateral petals* linear-oblong, c. 2.3 × 0.4 cm, sometimes the two petals fused. *Labellum* not deflexed, obovate, c. 2.5–2.8 × 1–1.4 cm, scarcely to deeply lobed, including the claw. *Staminodes* asymmetrically obovate with vein excentric, c. 1.3–1.5 × 0.5–0.6 cm. *Anther* thecae 6–6.5 × 2 mm, connective elongation angled to the pointed appendages, together 7–8 × 1–1.5 mm. *Ovary* hidden in leaf sheaths. *Epigynous glands* not seen. *Capsules* enclosed in leaf sheaths. *Seeds* brown, slightly constricted at middle, asymmetrically elliptic with apex apiculate, aril hardly lacerate.

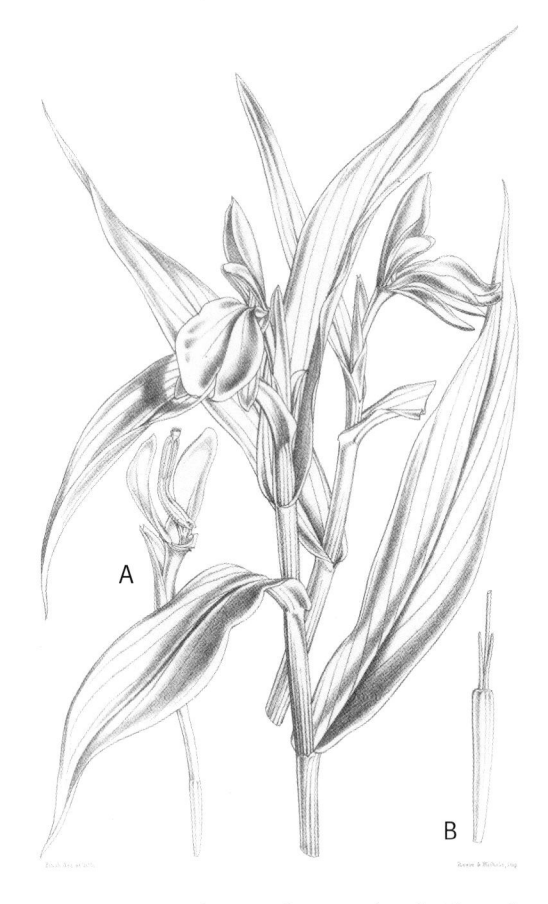

Fig. 44. Dissection diagram of **Roscoea brandisii** by Walter Hood Fitch. **A** tube of perianth with two lateral petals, stamen and stigma; **B** ovary and base of style.

10. ROSCOEA AUSTRALIS

This species of *Roscoea* may still have been growing, unknown and unsung, on Mount Victoria in West Central Burma, if it had not been for the vigilance of the explorer, professional plant collector and writer, Frank Kingdon Ward. He collected seed from that "bare" looking Burmese mountain in April 1956, when the heat had already completely dried up the steep grassy slopes. He was curious to know what had been growing there, and only by lying prostrate on the ground did he begin to see the dried up remains of the plants that had been so abundant, among them being dwarf irises, gentians and the roscoeas. Some of the gentians and the roscoeas had come adrift and even though they were in a skeletal state their capsules still retained seed. Kingdon Ward noted that as a general rule the roscoeas' soft capsules were held below the surface of the soil and were protected by the broad bases of their leaves.

Plate 9. *Roscoea australis*. KW 22124. Painted by Christabel King, July 1998.

THE GENUS ROSCOEA | **PLATE 9**

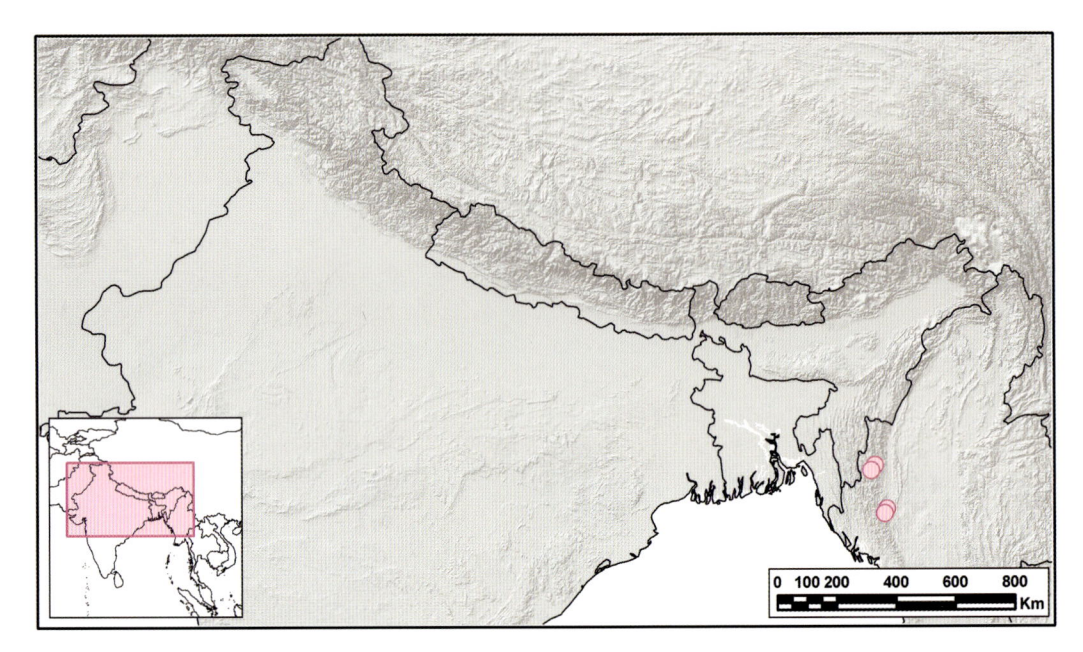

Map 13. Distribution of *Roscoea australis*.

This seed was eventually distributed under his collecting number, 22124, and produced the plants which are now in cultivation. That journey in 1956 was to be Kingdon Ward's last expedition to Burma, and his penultimate trip before his unexpected death in April 1958.

He went back to the same locality after the seasonal rains had been falling for five weeks during April and May. The slopes were emerald green with the grasses and "brilliant with flowers". In his book, *Pilgrimage for Plants*, published in 1960, he described the roscoeas as being a few inches high with one or two large deep purple flowers, opening in succession, and looking like orchids. He also noticed that where there were abundant swarms there were always several "milk-white" albino forms.

I described this species in 1982 from material from gardens, which must originally have received seed from Kingdon Ward. Kew has a herbarium specimen originating from plants cultivated at Calderstones Park, Liverpool in 1961; the flower colour was described as mauve. The specimen used for the description of the fruiting parts came from herbarium material at the Natural History Museum of the original wild-collected specimens. Flowers were preserved in spirit at Edinburgh by the late Rosemary M. Smith, a specialist in *Zingiberaceae*; the delicate flowers of gingers mean that such material is hugely valuable for detailed study. All these collections were recorded as originating from Kingdon Ward's collection 22124. When I described this species, I had seen only four other collections; one from the Calcutta herbarium, collected in the Chin Hills (*Dun* 85); a second, with purple flowers, at Kew, was collected 3.2 km (2 miles) from Haka at 2130 m in July 1910, (*Venning* 10); a third, collected from the type locality on Mt Victoria, came from the Edinburgh herbarium (*Cooper* 6009) and the fourth, also collected by Kingdon Ward (as KW 22292) was found by him in May 1956 at 2210 m on Mindat ridge; the specimen is at the Natural History Museum.

THE GENUS ROSCOEA 89
10. ROSCOEA AUSTRALIS

Since the original description, another Kingdon Ward specimen (KW 22380) from Mt Victoria (to 3053 m) has been seen at the Natural History Museum. Kingdon Ward collected it on the east ridge, between 2800 and 3000 m., on June 18 1956, and stated that the flowers were royal purple, sometimes white, and that the plants were "in their thousands on the south facing grassy slopes, burnt annually, same as (KW) 22124". The present political situation in Burma makes it unlikely that further collections will be possible in the near future. It can only be said that this species has only been found in the Chin Hills, a southern extension of the Sino-Himalayan plateau in west central Burma. However, Mt Victoria and Haka are about 150 km apart so the species may occur over quite a large area. The species grows in grassland, meadows and on mountain ridges at 2130–3000 m, and can be seen flowering from May to July.

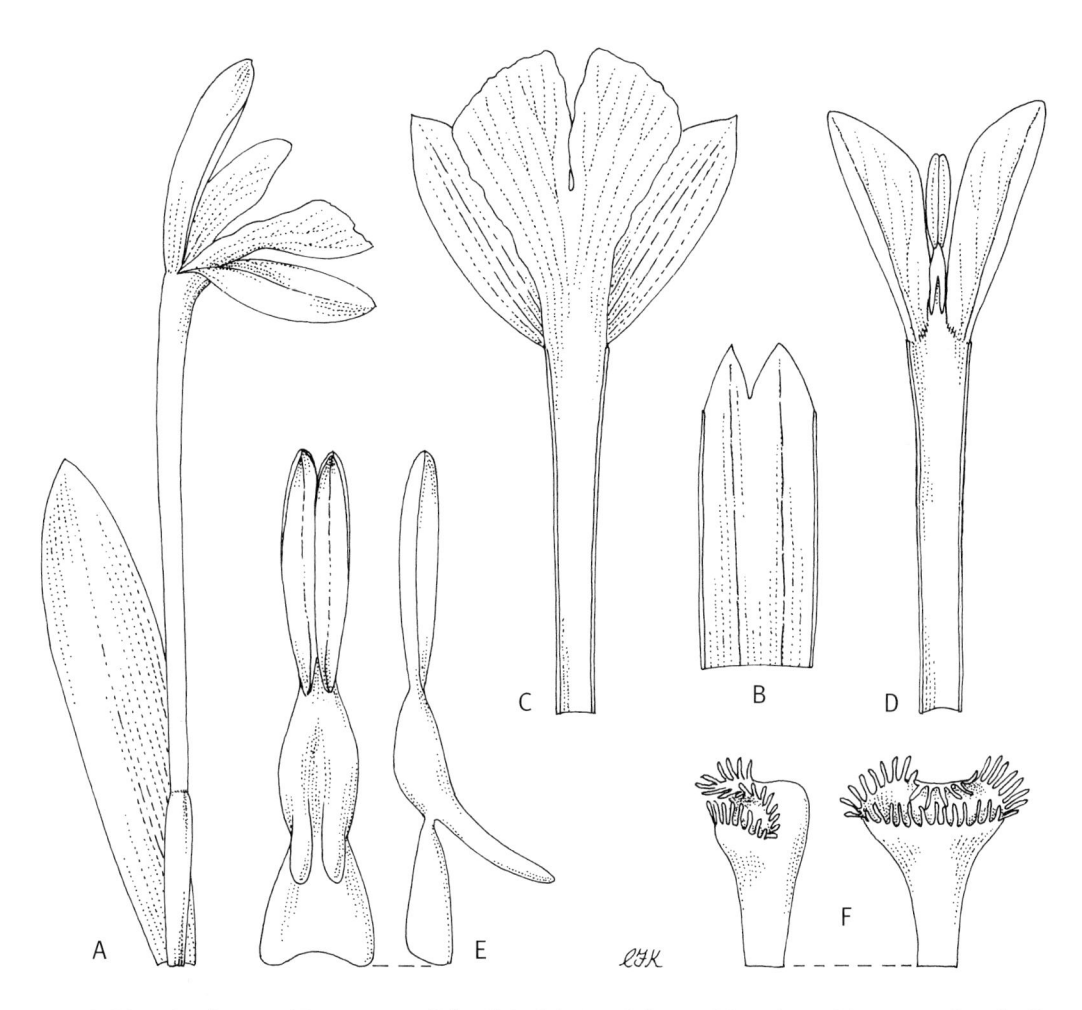

Fig. 45. Dissection diagram of *Roscoea australis* by Christabel King. **A** flower; **B** inner bract; **C** lower part of corolla; **D** upper part of corolla; **E** stamens, 2 views; **F** stigma, 2 views.

However, the type collection, *Kingdon Ward* 22124, was distributed widely, and the species is thriving in cultivation. The specimen used to prepare this plate was donated to Kew in 1996 by Mr Gary Dunlop from County Down, Northern Ireland. Mr Dunlop had wisely retained the Kingdon Ward collecting number with his plants, which he grows in the open ground. Photographs sent by him show this species at fruiting time with the stem swollen at about midway, showing the position of the capsules; this seems to contradict the observations made in the field by Kingdon Ward.

This species does not seem to have appeared in the horticultural literature until it is mentioned, in an article about the genus *Roscoea*, in a letter to the Editor of the *Gardener's Chronicle* that appeared in volume 167, on May 15 1970. Norman Hart of Uckfield, Sussex asked for information, saying "There is a roscoea under the number 2214 which is similar in stature to *R. alpina* but the purple flowers are held higher and therefore open fully". The Editor's note quite rightly observed that the number should be 22124, but then said, wrongly, that the plant is a form of *R. purpurea*. He added "It is more robust and taller than *alpina*". Unfortunately it is impossible to know if he was referring to the true *alpina*; *R. scillifolia* has often been referred to as 'alpina' in the literature.

When I wrote the original description, I had only seen herbarium specimens; these suggested that the corolla tube was always well exserted from the leaves. The few plants that I have seen in cultivation do not seem to show this trait. The three plants growing at Kew have flowers that sit just above the topmost leaves with hardly any tube showing, as can be seen in the plant illustrated in Plate 9.

Roscoea australis grows at 21°N in the tropical zone, and is isolated from other Indian species of the genus by 3–4 degrees latitude. Its specific epithet, *australis* (southern), was given because of this. However, it is geographically closest to *R. wardii*, which is found in eastern Tibet, Assam, northern Burma, and around the border with Yunnan province in China. In the herbarium and cultivated specimens of *R. wardii* that I have so far seen, the leaves are glaucous beneath, a character lacking in *R. australis*. There are also certain similarities with *R. tibetica*. However, the leaves of *R. tibetica* form a rosette, while *R. australis* has them arranged distichously.

Roscoea australis Cowley in Kew Bull. 36(4): 770–771 (1982).
Type: West Burma, Mt Victoria, 2800 m, fr. April, 1956, *Kingdon-Ward* 22124 cult. Calderstones Park, Liverpool, April 1961 (holotype K; isotype BM (fr.); spirit material of flowers from living specimen at E).

DESCRIPTION. *Rhizomatous, tuberous rooted herb* to 30 cm. *Sheathing leaves* 2–3, obtuse, whitish, speckled, soon splitting. *Leaves* 2–7, oblong-ovate to lanceolate, 4–17 × 1.5–4 cm, distichous, widest at the base, slightly auriculate, acute to acuminate, glabrous, shining green above, lighter below, side veins parallel, ligule area ± semicircular. *Ligule* raised 1.5 mm. *Inflorescence* not pedunculate, flowers 1, occasionally 2, open at a time, purple, mauve-pink or white. *Bracts* shorter than the calyx, c. 3.3 × 1.1 cm, linear, obtuse to truncate, emarginate. *Calyx* c. 4 cm long, bluntly tridentate, spotted pinkish brown, darkening towards the apex. *Corolla tube* sometimes long-exserted, white. *Dorsal petal* obovate, 2.2–3 × 1.7–2 cm, without a claw, strongly cucullate, apiculate. *Lateral petals* oblong, 1.9–2.6 × 0.6–1 cm, with whitish base, obtuse. *Labellum* ± deflexed, obovate, 2–3 × 1.9–2.4 cm including c. 4 mm claw, bilobed to ± half way, each lobe emarginate. *Staminodes* asymmetrically obovate with vein excentric, 1.3–2.2 × 0.6–0.9 cm. *Anther* white; thecae 5–6 × 2 mm; connective elongation angled to the greenish, obtuse appendages, together 6–7 × 2.5 mm. *Epigynous glands* and *ovary* not seen. *Seeds* dark brown, asymmetrically elliptic with slight constriction at the middle, aril shallowly lacerate.

11. ROSCOEA WARDII

We know nothing of the origins of this plant, which has been growing at Kew since 1964. The best guess is that it was introduced to cultivation by Frank Kingdon Ward. Of the wild collections of this species in the herbarium, the vast majority were collected by him. Unfortunately it is often true, as with Ward's collections of *Roscoea australis* from Mount Victoria in Burma, that the collecting data are lost when plants are distributed; growers in the past often seem to have failed to retain the information relating to their plants.

Kingdon Ward collected this species many times when in the monsoon belt in the area of the Upper Irrawaddy, in the boundary areas of Tibet, Assam, Burma and China, between the years 1926 and 1933, and he was struck by the "flowers of rich tyrian purple" when collecting in the Adung Valley in July 1931 (KW 9682). However, in his many accounts of those trips in his books and articles, he never mentions these Roscoeas. We have to rely on the notes that he wrote on the labels accompanying his pressed collections. Luckily, these notes often go into great detail, so that they form a valuable resource for the researcher. Because of this, I used his name when choosing a specific epithet for this species. This species is without doubt in cultivation in many gardens that are our horticultural treasure houses.

When I described this species, I compared it with *Roscoea tibetica* from further north-east in Tibet and eastwards in the Yunnan and Sichuan Provinces of China, close to the Burma borders, and with *R. australis*, from further south in Burma. *R. wardii* differs from both these species in producing only 2–3 leaves, which are glaucous beneath. No other species of *Roscoea* is known from this area, which is far removed from other known localities for the genus. The flowers of *R. wardii* are large, and the labellum, which has emarginate lobes that give it a frilled appearance, is very distinctive. This, and the deep colour of the flowers make it a very attractive species in cultivation. Handel-Mazzetti,

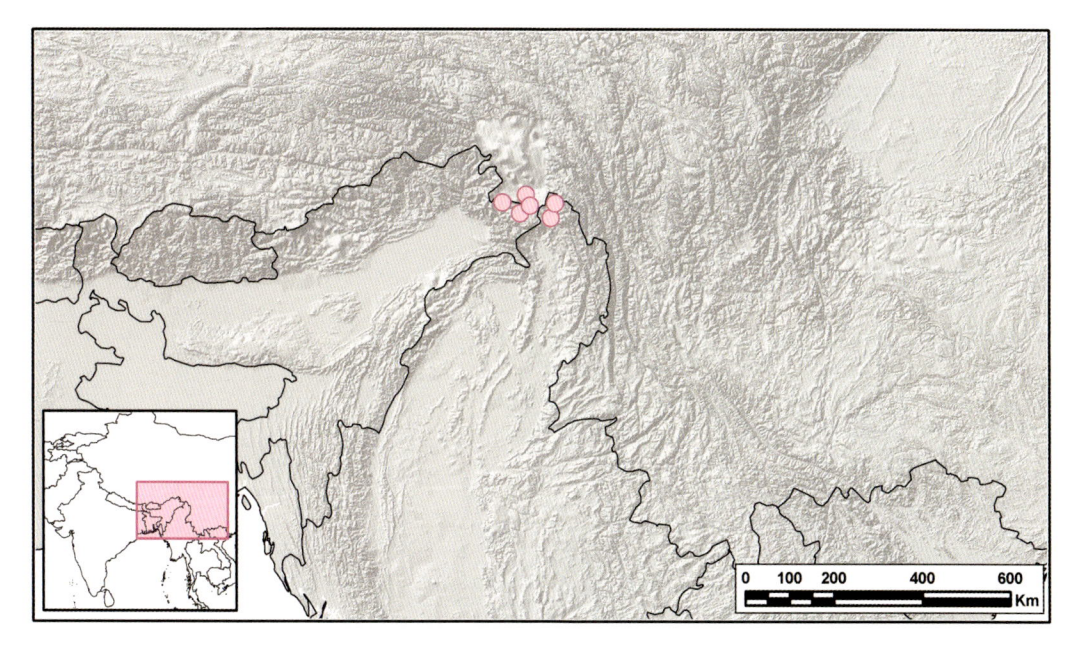

Map 14. Distribution of **Roscoea wardii**.

92 | THE GENUS ROSCOEA
PLATE 10

in his description of *R. blanda* var. *pumila*, a synonym of *R. wardii*, describes the colour of the flowers of one of his collections (3351) as rose and the type specimen (9510) as purple. The pink form is probably not in cultivation.

Since this species was described, a new species from Bhutan has been described to include specimens previously included in the extremely variable *Roscoea tibetica*. Bhutan has been recently become more accessible to scientists and eco-tourism, and botanists from the Royal Botanic Garden, Edinburgh have been able to study plants in the wild towards their publication of the *Flora of Bhutan*. Better herbarium material has been collected, allowing the description of the new species, *R. bhutanica* Ngamriab. (Ngamriabsakul & Newman 2000). The two species are clearly similar, but there are numerous differences between them. The labellum in *R. wardii* is clawed, but not in *R. bhutanica*. The staminodes of *R. wardii* have a shorter claw than in *R. bhutanica*, and the first bract is closed and envelops the further bracts in *R. wardii* but not in *R. bhutanica*. There are more leaves per plant in *R. bhutanica* and the connective elongation of the anther is longer in *R. bhutanica*. The labellum shape also appears to be different, although it should be said that in some species, such as *R. purpurea*, the labellum can vary in shape both within and between populations.

At maturity the smooth flowering stalk of *Roscoea wardii* becomes shortly exserted from the leaves but can be distinguished from those forms of *R. cautleyoides* with a similar habit by the fewer, shorter, wider leaves, and by the bilobed calyx with broad and obtuse lobes (not bidentate and acute). The corolla tube of *R. wardii* is usually well exserted from the calyx, and it has longer staminodes and anthers than has *R. cautleyoides*.

Roscoea wardii occurs in eastern Tibet; Kingdon Ward's collection came from Zayul. He also collected the species in Assam, in the Delei valley, and at the frontier in the Di Chu valley. His specimens from Burma came from the Adung valley. *Farrer* 1766 is a collection from the Burma-Yunnan borders; *Yu* collected specimens in Yunnan itself near the borders in the upper Kiukiang valley.

The species grows in these areas in meadows and open, grassy places, along margins of dense thickets, shady and moist places, and beneath bushes in *Abies* or *Rhododendron* forest. They can also be found on south facing gravel and alpine turf slopes, or on mud slides recently uncovered by melting snow at altitudes between 2440 and 3960 m. Flowering is from May to the beginning of August.

Roscoea wardii Cowley in Kew Bull. 36(4): 768–770 (1982). Type: India, Assam, The Chu valley, 2740 m, July 1950, *Kingdon Ward* 19623 (holotype BM).

R. blanda K. Schum. var. *pumila* Hand.-Mazz. in Symbol. Sinicae 7: 1321 (1936). Type: China, Yunnan, Tibet/Burma borders between Nu Jiang and Irrawaddy rivers, 3500–3800 m, July 1916, *Handel-Mazzetti* 9510 (holotype WU).

DESCRIPTION. *Rhizomatous, tuberous rooted herb* to 32 cm. *Sheathing leaves* 3–5 with pink veins. *Leaves* 2–3, elliptic, 7–8 × 1.7–4.5 cm, apex acute to acuminate, glaucous below, hyaline edge minutely papillate, side veins parallel. *Ligule* area ± semicircular, ligule raised 1–2 mm. *Inflorescence* shortly or hardly pedunculate, peduncle smooth, flowers rose or deep purple. *Bracts* 3.5–5.5 × 1–1.4 cm, shorter than the calyx, exserted from the leaves, obtuse, ciliate, pale, first and second bract fused at first, then splitting when second flower opens. *Calyx* obtuse, bilobed, 3.3–4.5 cm, lobes broad,

---

Plate 10. *Roscoea wardii*. Cultivated at R.B.G., Kew. Painted by Christabel King, June 1999.

rounded, exserted, ciliate,. *Corolla tube* 4–8 cm, usually exserted from calyx at maturity. *Dorsal petal* obovate or broadly elliptic, 2–3.2 × 1.3–2.5 cm, not clawed. *Lateral petals* linear oblong to oblong, 1.7–3.4 × 4.5–14 mm. *Labellum* deflexed, obovate, 2.2–4.5 × 1.6–4.5 cm, including 8–10 mm claw, deeply lobed, each lobe emarginate, with three white raised stripes at the base of each lobe. *Staminodes* elliptic, 1.6–2 × 0.7–1 cm, with ± median vein, shortly clawed. *Anther* white, angled below the thecae, thecae 6–9 × 2–3 mm, connective elongation in line with the yellow, obtuse appendages, together 5–8 × 2–3 mm. *Ovary* 10–12 × 2.5– 4 mm. *Epigynous glands* 4 mm long. *Style* purple. *Stigma* white. *Capsule* c. 3 × 1 cm. *Seeds* rugose, elliptic to triangular with constriction near the middle, aril shallowly lacerate.

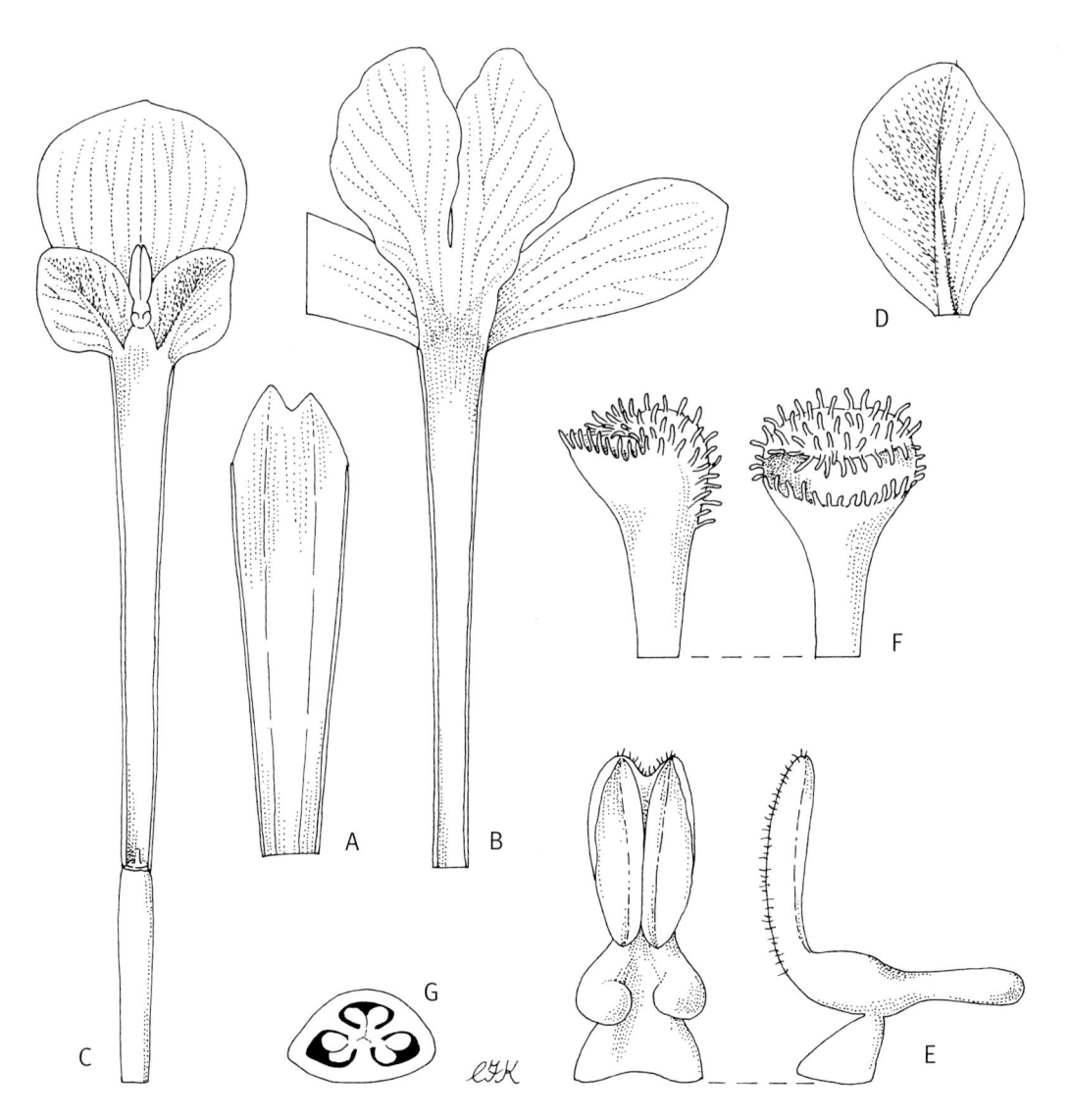

Fig. 46. Dissection diagram of **Roscoea wardii** by Christabel King. **A** floral bract; **B** lower part of corolla; **C** upper part of corolla; **D** inner perianth segment; **E** stamens, 2 views; **F** stigma, 2 views; **G** ovary, transverse section.

THE GENUS ROSCOEA
11. ROSCOEA WARDII

Fig. 47. ***Roscoea wardii*** cultivated at R.B.G., Kew. Photographed by Richard Wilford, R.B.G., Kew.

12. ROSCOEA SCHNEIDERIANA

This striking ginger, which has recently been reintroduced to cultivation, was first described as a variety of *Roscoea yunnanensis*. Although that species is now regarded as conspecific with *R. cautleyoides*, its variety, *schneideriana*, is specifically distinct. It has, furthermore, three features that make it unique within the genus.

It differs from *Roscoea cautleyoides* in that the inflorescence is not borne on a significantly elongated peduncle. The staminodes of *R. schneideriana* are almost as long as the dorsal petal; in *R. cautleyoides* they are much shorter. Another variety of *R. yunnanensis*, var. *dielsiana*, which was described at the same time as *schneideriana*, is inseparable from the latter variety and appears to be just a narrow-leaved variant. Young cultivated plants of *R. schneideriana* grown from seed produce much narrower leaves than more mature plants. Plants may also flower in cultivation as early as their second year. In most species of *Roscoea*, the dorsal petal is the same colour as the lateral petals, but in *R. schneideriana* the dorsal petal is a speckled, dusky purplish pink while the lateral petals, staminodes and labellum are deep violet purple. *R. schneideriana* has a hooked stigma; in all other species it is erect and helmet-shaped. The ends of the anther appendages are white and rounded and protrude just at the base of the labellum, another feature unique to this species (Fig. 49). There is also a small petaloid flap attached to the lower half of the straight edge of the staminode which is then free from the staminode for about a third of its width (see Fig. 48). This may not occur in every individual; it is a feature that is very difficult to see on herbarium specimens. The mature leaves are arranged in an unusual fan-shape; such an arrangement does not occur in any other species.

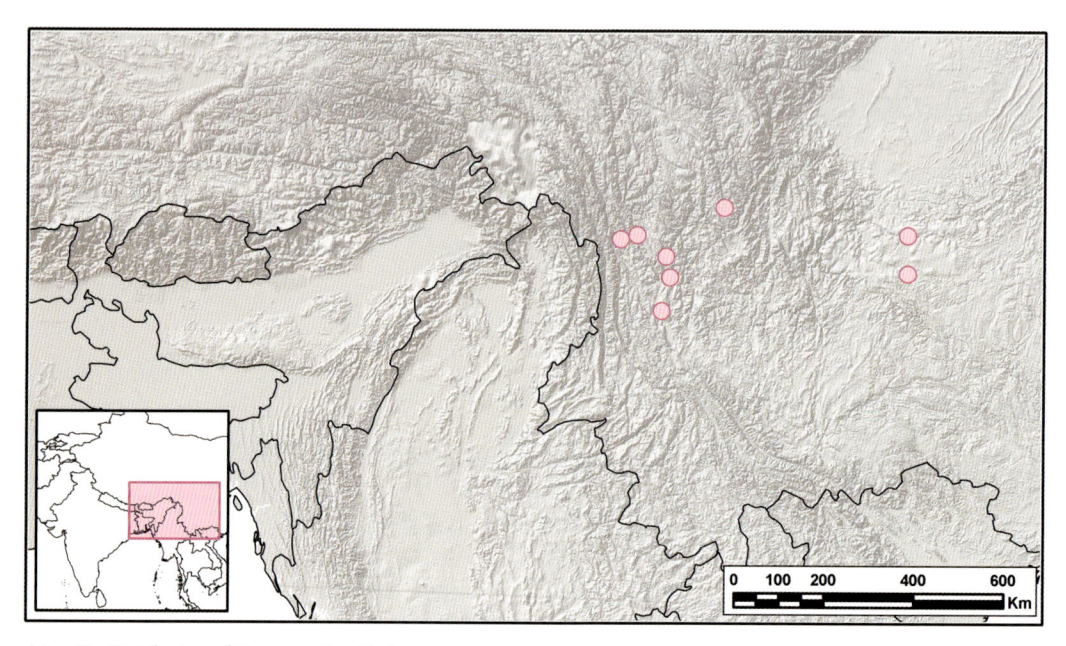

Map 15. Distribution of **Roscoea schneideriana**.

Plate 11. *Roscoea schneideriana*. CLDX 0686. Painted by Mark Fothergill, July 1993.

THE GENUS ROSCOEA
PLATE 11

98 THE GENUS ROSCOEA
12. ROSCOEA SCHNEIDERIANA

The plant illustrated here came from a population that I found in October 1990 in Yunnan, in woodland dominated by *Pinus armandii* and *P. yunnanensis*. The purple-flowered form of *Roscoea cautleyoides* grew with it. A few non-flowering plants and seeds were collected (CLD 0686), and these flowered in cultivation at Kew in July 1992. Other collections of the same species were made on the Chungtien-Lijiang-Dali Expedition (CLD 0773, 0774B and 1238 (Fig.

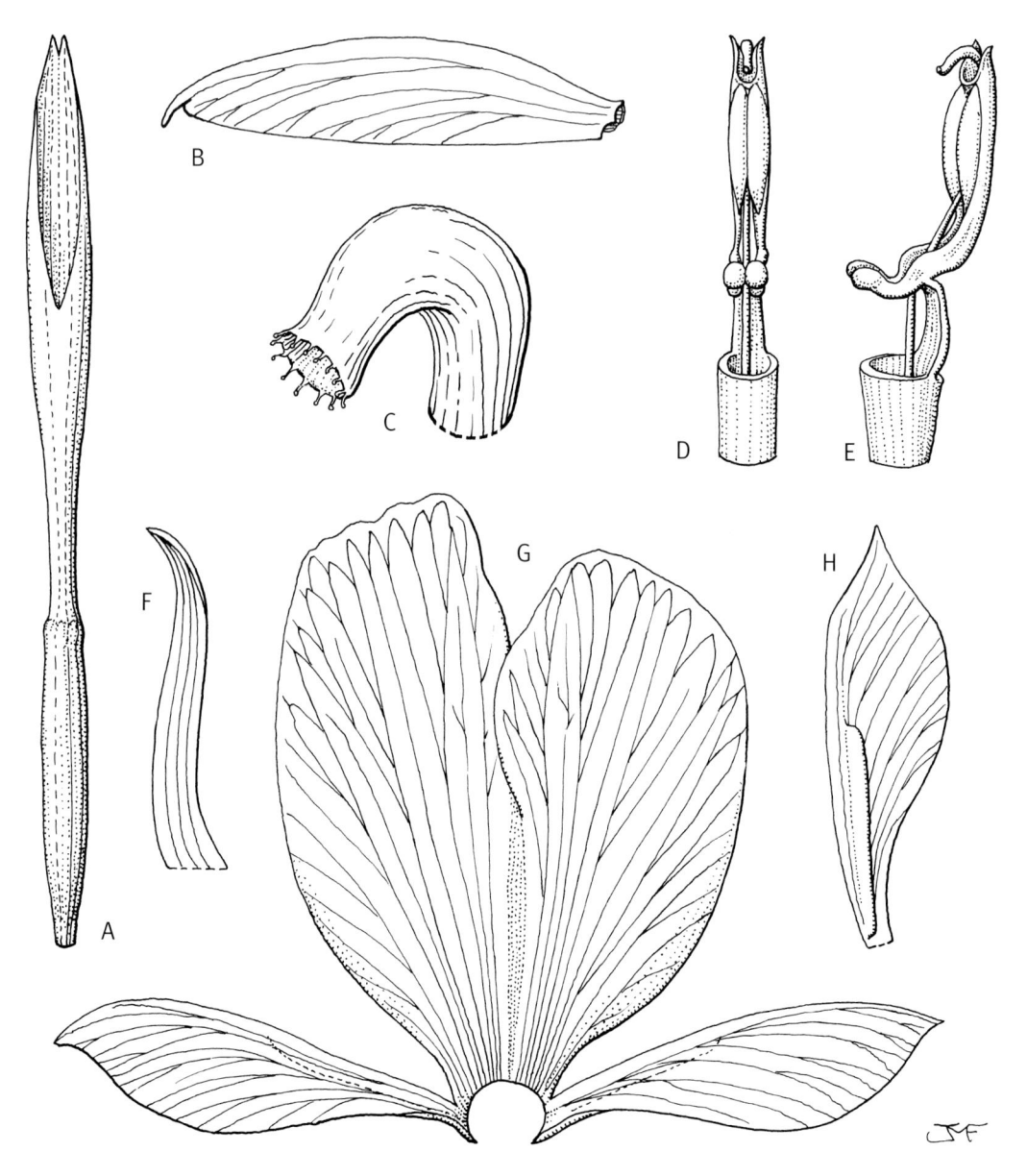

Fig. 48. Dissection diagram of **Roscoea schneideriana** by Mark Fothergill. **A** ovary and calyx; **B** dorsal petal; **C** stigma; **D** stamen, front view; **E** stamen, side view; **F** lateral petal; **G** labellum and staminodes; **H** staminode showing phlange.

49). The first of these flowered at Kew in July 1991, but CLD 0686 has proved to be the collection with most horticultural merit, because the large flowers protrude well beyond the leaves. The collections all came from the Yulong Shan mountain range in Yunnan province, near Lijiang, the actual plant figured originating from Wen Bi Shan south-west of Lijiang at an altitude of 2800 m. Subsequent collections came from the Gang Ho Ba, an old river valley, on steep rocky slopes below pine trees at 3150 m and close by at the roadside in open situations. The last collection mentioned above (CLD 1238) was growing at the north end of the Lijiang Plain on limestone rocks at 2900 m.

A later visit to the Gang Ho Ba sites near Lijiang in June 1992 gave some cause for concern. Species such as this one, whose habitat seems to be fairly restricted, are vulnerable to disturbance and to changes in their environment. Between the autumn of 1990, when this species was originally collected, and the visit in June 1992, areas of pines had been felled, reducing the area of habitat suitable for this species. No trace of *Roscoea schneideriana* could be found at the locality where it had formerly grown with *R. humeana* and *R. cautleyoides* (Fig. 13). However, it may just be that the new shoots had not yet appeared. Studies of herbarium material suggest that *R. schneideriana* starts flowering later than *R. cautleyoides*, and the latter species was still present, although not so abundantly, in spite of the removal of the shade of the pines. Nevertheless, there may still be a threat to its survival in this area, and so the maintenance of material in cultivation hopefully provides some degree of insurance for the future, and represents a positive step towards its conservation.

Fig. 49. **Roscoea schneideriana**, CLDX 1238, collected in Yunnan province, China, in cultivation in the Alpine House at R.B.G., Kew. Photographed by Andrew McRobb, R.B.G., Kew.

12. ROSCOEA SCHNEIDERIANA

George Forrest may have introduced *Roscoea schneideriana* to cultivation from one of his many forays into western China at the beginning of the last century. Cowan (1938) includes a photograph of a plant labelled *R. purpurea*. The photograph shows quite clearly the distinctive fan-shaped leaf arrangement and the tell-tale round white ends to the anther appendages, both showing clearly that it depicts *R. schneideriana*. The photograph appeared again (p. 220) in an article on *Roscoea* in *The Journeys and Plant Introductions of George Forrest* in 1952, edited by the same author. This time the photograph is labelled as having been taken by Forrest, although it would seem likely that Forrest was the collector and Adam the photographer, as originally stated. It seems likely from the arrangement of the plants that the photograph was taken in the Edinburgh Botanic Garden, and not in the wild.

Camillo Karl Schneider, after whom this species is named, was a prolific collector of herbarium specimens who travelled in south-eastern Europe in 1907 and 1908 and in western China in 1913 and 1914.

Roscoea schneideriana grows in the southern Sichuan and Yunnan Provinces of western China. The habitat it favours is shady situations in mixed forest, open limestone slopes, moist stony pastures, among boulders, between rocks and on ledges of mountain cliffs at altitudes between 2600 and 3350 m; it flowers in July and August.

Fig. 50. *Roscoea schneideriana*. Cultivated and photographed by John Fielding, Sheen.

Roscoea schneideriana (Loes.) Cowley in Kew Bull. 36(4): 762 (1982).
Type: China, Yunnan, on limestone slopes near Lijiang town, 2800 m, July 1914, *Schneider* 1770 (holotype B, presumed destroyed; isotypes G, K, US).
R. *yunnanensis* Loes. var. *schneideriana* Loes. in Notizbl. Bot. Gart. Berlin-Dahlem 8: 600 (1923). Type as above.
R. *yunnanensis* var. *dielsiana* Loes., *loc. cit.* (1923). Types: China, Yunnan, near Pé-long-tsin, 3200 m, *Bonati* Sér. B, 3462 (*leg.* Maire) (US); near Kin-tschong-schan, 2600 m, *Bonati* Sér. B, 2989 (*leg.* Maire) (not seen).

DESCRIPTION. *Rhizomatous, tuberous rooted herb* (9–)28–38(–44) cm tall. *Sheathing leaves* 3–4, obtuse. *Leaves* (2–)4–7, distichous or forming a rosette, the first true leaf ovate, subsequent ones linear to narrowly lanceolate, (5–)10–22 × 0.4–2 cm, acute to acuminate, glabrous; *ligule* area more or less semicircular, ligule 0.5 mm high. *Inflorescence* eventually with shortly exserted peduncle, or not. *Bracts* 3.3–7 cm long, 0.3–1.4 cm wide, elliptic, acute to narrowly acute, the lowest tubular at base. *Flowers* rich dark purple, pinkish purple or white, with differently-coloured dorsal petal, usually one flower open at a time; *calyx* 3–4 cm long, sharply but shallowly bidentate, acute; *perianth tube* 4–5.5 cm long, scarcely exserted from calyx; *dorsal petal* elliptic, pinkish purple, speckled, 1.7–3.5 × 0.25–1.2 cm, narrowly acute, apiculate; *lateral petals* linear-oblong, acute, 2–3.5 × 0.3–0.5 cm; *labellum* obovate, usually lobed for half its length, each lobe emarginate, 1.8–3.5 × 1–2.5 cm. *Staminodes* asymmetrically obovate to rhombic, 1.5–2.5 × 0.4–0.7 cm, almost as long as dorsal petal, with a weak vein and sometimes with a small flange attached. *Anther* white, sometimes crested, with thecae 6–9 × 1–1.5 mm; connective elongation angled to the obtuse white appendages which have rounded ends, together 8–9 mm long. *Epigynous glands* c. 5 mm long. *Stigma* gradually widening, abruptly uncinate, narrowly infundibuliform. *Seeds* dull brown, more or less turbinate, arils lacerate.

13. ROSCOEA SCILLIFOLIA

This species has been known for many years in the horticultural trade in its pink form as 'alpina'. It may well be endangered in the wild, because it is known only from a limited area where recent efforts to find it growing have failed.

The exact collecting data for the two colour forms depicted here is not known, but, according to an article written by Cowan (1938), the original wild stock from which the cultivated plants derive was collected in China by Forrest. There are certainly herbarium specimens in existence collected by him in which the two colour forms are mounted on a single sheet, suggesting that they came from the same population. Cowan himself was familiar with Himalayan plants as he had worked for the Indian Forest Service and collected in Sikkim, Bengal and Burma. He had obviously seen many herbarium specimens of *Roscoea*, and must have been aware that this narrow-leaved plant with small pink flowers was not the same as the familiar *R. alpina* found from Kashmir eastwards along the Himalaya to Bhutan and into Tibet. He asks the reader: "What are the plants now in gardens under the name *R. alpina* raised from seed collected by Forrest on the Lijiang Range at 8 – 12,000 feet ………?" A photograph of the plant under discussion, presumably growing in the Edinburgh Botanic Garden, is featured opposite the text.

The herbarium specimens used to describe what is now known as *Roscoea scillifolia* were collected in 1887 and 1888 by Delavay from Yunnan Province in China, at Hee-chan-men near Dali, a popular destination for present-day tourists. Gagnepain described them as a variety of the Nepalese *R. capitata*. The type material included three forms, white, rose and the deep purple one, which Delavay described as "red (rouges)". The first two forms he numbered 2685 and the purple form he called 2685 bis to distinguish them. Gagnepain favoured the name "scillifolia" as he said the leaves resembled those of *Scilla bifolia*.

In 1923, Loesener transferred the varietal epithet "*scillifolia*" to his species *Roscoea yunnanensis*, which is a synonym of *R. cautleyoides*. Among the herbarium specimens that he cited under *R. yunnanensis* var. *scillifolia* is *Forrest* 4809; this is a mistake, for it is in fact *R. cautleyoides*.

In cultivation, the two colour forms flower at different times although not consistently one before the other. Records show that in Edinburgh Botanic Garden the pink form flowers before the purple one, but the reverse is true at Kew.

102 | THE GENUS ROSCOEA
PLATE 12

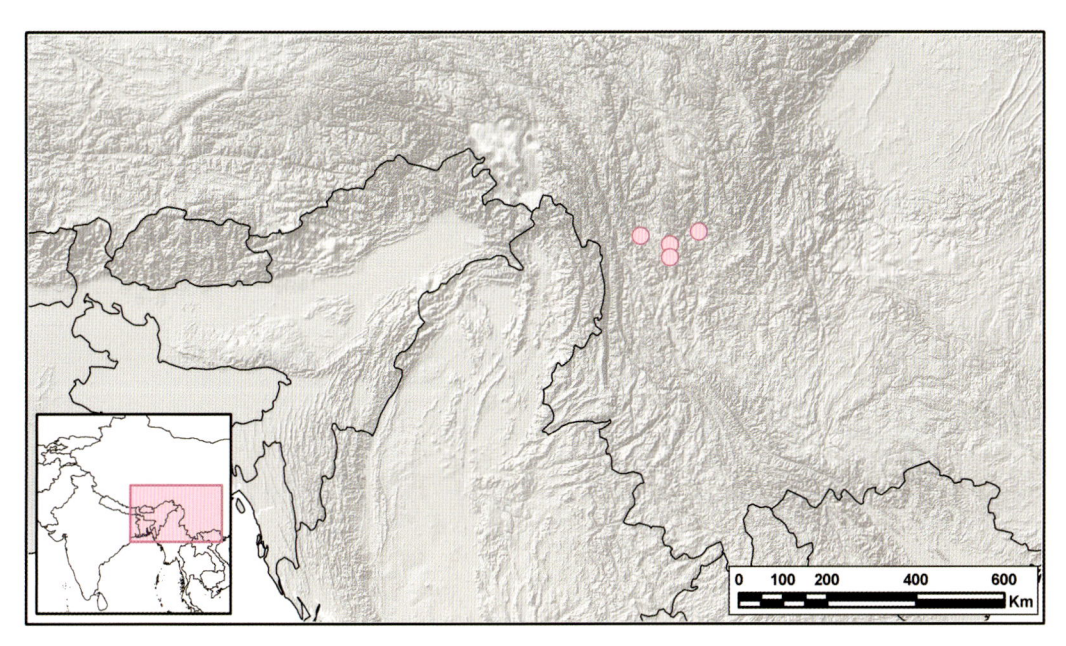

Map 16. Distribution of **Roscoea scillifolia**.

Cowan (1938) described variation within the pink form; sometimes the plants are squat with wide leaves, and sometimes tall with narrower leaves. The inflorescence may have only a short peduncle so that the flowers are not exserted above the leaves, or a long peduncle that holds the flowers well above the leaves (Fig. 52). The purple flowered form is less diverse, and is usually tall with long-exserted peduncles and narrower leaves (Fig. 53). The bracts surrounding the capsules of the pink form tend to be wider than those of the purple form. However, in populations of both colour forms, there are plants that show a wide range of variation in all their characters. Herbarium specimens (*Forrest* 6354) of a purplish-blue flowered form, from the eastern flank of the Lijiang range in Yunnan, show plants with a compact, wide-bracteate inflorescence like that of the cultivated pink form. However, the small flowers are more or less constant in size and shape in both forms.

The two forms illustrated here came from cultivated stock. The purple form is described below as *Roscoea scillifolia* forma *atropurpurea*. The pink form, which becomes *scillifolia* forma *scillifolia* was donated by the Dix and Zijerveld Bulb Nursery in Holland in 1993, and forma *atropurpurea* came from Edinburgh Botanic Garden in 1977 under the name *R. yunnanensis*.

The many references to this species in the horticultural literature may refer either to the present species or to *Roscoea alpina*. The articles in the *Gardener's Chronicle* by H.E. Bawden in 1960 (Vol. 147); by R.F. Watson in 1961 (Vol. 150); by W. Ingwersen in 1963 (Vols. 153 & 154) and by Norman Hart in 1970 (Vol. 167) all use the name *R. alpina*, but all of them really refer to *R. scillifolia*.

Plate 12. *Roscoea scillifolia* forma **scillifolia** (left) and forma **atropurpurea** (right), painted in August 1998 and June 1998 respectively, by Christabel King.

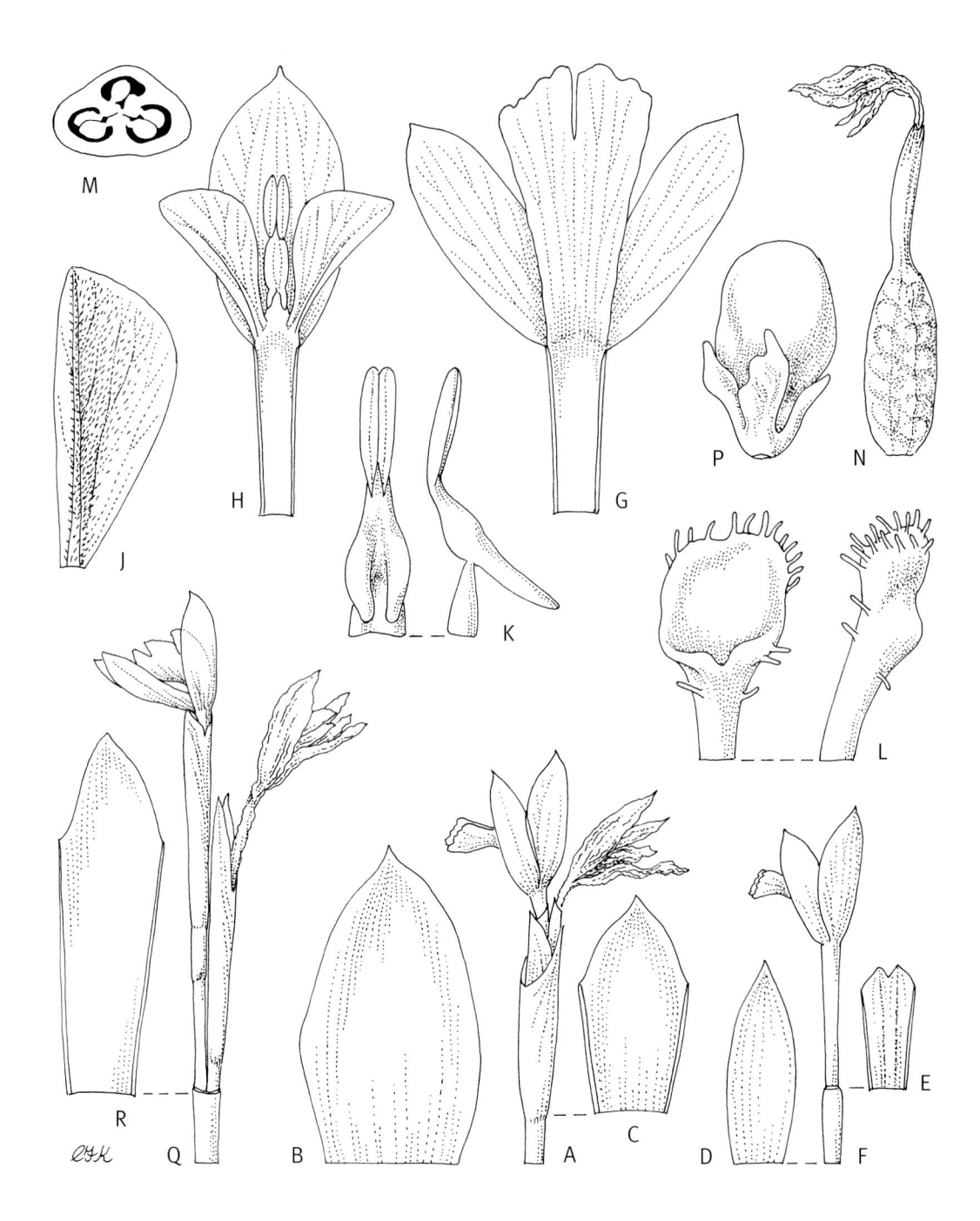

Fig. 51. Dissection diagram of *Roscoea scillifolia* by Christabel King. **A**–**P** Pink form: **A** inflorescence; **B** & **C** inflorescence bracts; **D** outer floral bract; **E** inner floral bract; **F** flower; **G** lower part of corolla; **H** upper part of corolla; **J** inner perianth segment; **K** stamens, 2 views; **L** stigma, 2 views; **M** ovary, transverse section; **N** ripe capsule; **P** seed. **Q** & **R** Dark form: **Q** inflorescence of 2 flowers; **R** inflorescence bract.

This species is not one of the showiest members of the genus and fairly disparaging remarks have been made about it. Again in the *Gardener's Chronicle* (May 15 1970, Vol. 167), Norman Hart comments "*Roscoea alpina* (*scillifolia*) is a quite miserable plant. It seems almost ashamed of its flowers and keeps them nearly hidden, low down in the leaf sheath, so that they barely have room to open fully........."

This species could be extinct in the wild. The last known collections were made at the beginning of the 20th Century, in mountains to the east of the Yangtze river loop and further south, in the Yulong Shan area around Lijiang. Plants grew in open, stony, moist mountain pastures between 2740 and 3350 m, and flowered from June to August.

Roscoea scillifolia (Gagnep.) Cowley in Kew Bull. 36(4): 765–766 (1982).
R. capitata Sm. var. *scillifolia* Gagnep. in Bull. Soc. Bot. France 48: LXXIV (1901, published in 1902). Type: China, Yunnan, Hee-chan-men, 2800 m, June 1888, *Delavay* 3283 (lectotype P; isotypes CAL, K). Original syntypes: Hee-chan-men, Kan-hay-tze, June 1887, *Delavay* 2685, 2685 bis (P).
R. yunnanensis Loes. var. *scillifolia* (Gagnep.) Loes. in Notizbl. Bot. Gart. Berlin-Dahlem 8: 600 (1923).
[*R. alpina sensu* auct. hort.; see list above].

Fig. 52. ***Roscoea scillifolia*** forma ***scillifolia*** cultivated in the Rock Garden at R.B.G., Edinburgh. Photographed by Richard Wilford, R.B.G., Kew.

13. ROSCOEA SCILLIFOLIA

ILLUSTRATION. *The Rock Garden* 26(4): fig. 116; text p. 323 (2000).

DESCRIPTION. *Plants* (6–)10–27(–37) cm tall. *Sheathing leaves* up to 3, obtuse, soon splitting, surrounded by brown collar of old leaf sheaths. *Leaves* 1–5 including underdeveloped ones, linear to lanceolate, (6–)11–21.5(–25) × (0.6–)1.5– 2(–2.7) cm, the lower sometimes falcate, acute to obtuse, surface pustulate above, punctate below, apex scabrid; side veins parallel. *Ligule* area ± semicircular in compact plants, triangular in mature leaves of plants with pedunculate inflorescences; ligule raised 2–3 mm. *Inflorescence* with or without exserted, slightly ribbed peduncle; flowers deep blackish purple, pink, white or occasionally mauve; one flower open at a time. *Bracts* green, ± equal to or longer than the calyx, 2.6–5 × 1–2.3 cm, acute to obtuse, ciliate; first tubular bract, enclosing inflorescence, soon splitting. *Calyx* 1.5–2.1 cm, whitish brown, apex narrow, bluntly bidentate, occasionally tridentate. *Corolla tube* hardly exserted from calyx at maturity, 1.6–3 cm long, shorter than the bract. *Dorsal petal* elliptic, 1.4–2 × 0.6–1 cm. *Lateral petals* linear oblong 1.1–2 × 0.4–0.65 cm. *Labellum* 1.3–2 × 0.8–1.2 cm, obovate, scarcely to deeply lobed. sometimes each lobe emarginate, with white lines at throat. *Staminodes* elliptic to asymmetrically obovate, 1–1.4 × 0.3–0.5 cm, vein excentric, papillose, shortly clawed. *Anther* white, angled below the thecae; thecae 4–5.5 × 1.5–2 mm; connective elongation in line with the pointed appendages, together 5–6 × 1.5–2 mm, swollen at joint with filament. *Epigynous glands* 2.7–4.1 mm. *Ovary* 1–1.5 × 0.3–0.45 cm, triangular in section. *Style & stigma* white. *Seeds* elliptic to triangular; aril shallowly lacerate.

Fig. 53. ***Roscoea scillifolia*** forma ***atropurpurea*** cultivated at R.B.G., Kew. Photographed by Richard Wilford, R.B.G., Kew.

Roscoea scillifolia (Gagnep.) Cowley forma **atropurpurea** Cowley, forma nov. a forma *scillifolia* floribus atropurpureis nec roseis differt. Typus: China, Hee-chan-men, Kan-hay-tze, June 1887, *Delavay* 2685 bis (holotypus P).

DESCRIPTION. Differs from the typical form (forma *scillifolia*) by having dark purple flowers rather than pink and white.

14. ROSCOEA CAUTLEYOIDES

This attractive plant was described by François Gagnepain in 1902, from collections gathered at the end of the 19th Century by Abbé Jean Marie Delavay and Prince Henry d'Orléans. Once it had been introduced into cultivation, it quickly became well known and popular among horticulturalists. George Forrest, who went on expeditions to western China on behalf of Arthur K. Bulley and his seed firm of Bees', collected this species many times at the beginning of the twentieth century, as did many other collectors. The plant illustrated in the *Botanical Magazine* (t. 9084) in 1926 was a Forrest collection (2070) of the yellow form of this species. That same year, a plant described as a pale sulphur-yellow form, *Roscoea cautleyoides* 'Beesii' was submitted by Bulley to the Royal Horticultural Society, and received an Award of Merit (*Journal of the Royal Horticultural Society* 52: lvi). The description suggests that it could be the same form that is now called 'Jeffrey Thomas' (Fig. 99). *R. cautleyoides* had itself received the same award in 1913, again given to a plant submitted by Bees' (*Journal of the Royal Horticultural Society* 39: cxxxiii–iv).

The illustration of the purple form of this species (Plate 14), was prepared by Christabel King, and shows plants collected on the Chungtien-Lijiang-Dali Expedition in 1990 (collection number CLDX 0687), and cultivated at Kew (Fig. 55). The plants were collected in China in Yunnan Province near Lijiang at Wen Bie Shan, at the edge of pine woodland at 2800 m on 3 October when the plants had finished flowering. When I visited the same area two years later, I was interested to see populations of yellow-flowered plants including occasional purple, dusky pink or white-flowered individuals; populations of purple-flowered plants likewise included occasional individuals of the other colour forms. A single photograph (Fig. 56) exists of a plant with rust-coloured flowers, apparently intermediate between the yellow and purple-flowered forms.

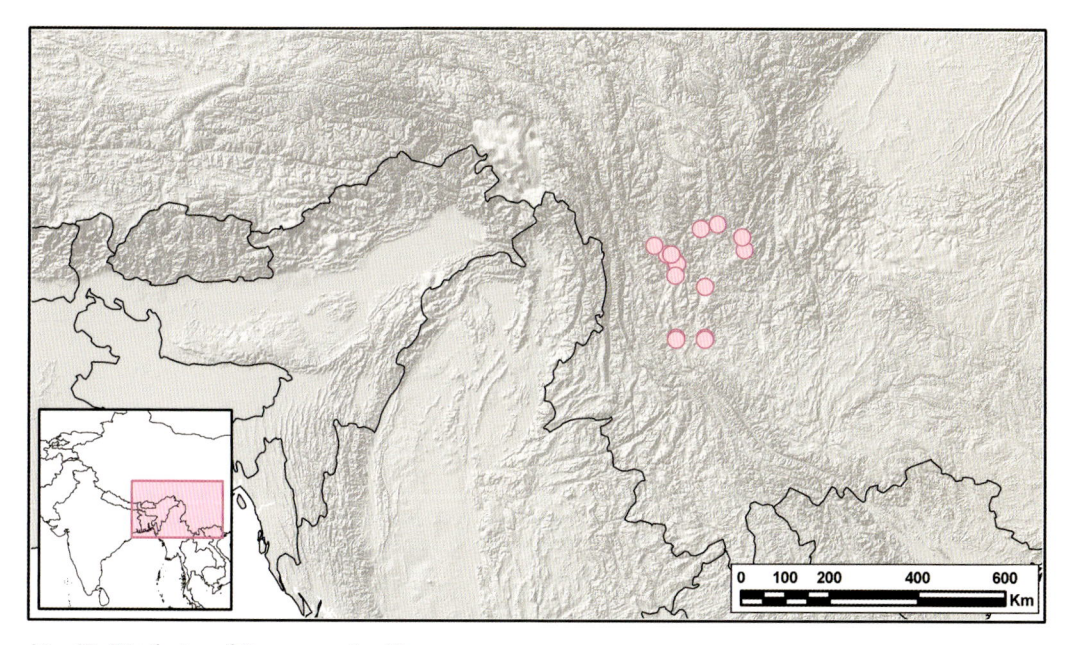

Map 17. Distribution of ***Roscoea cautleyoides***.

108 THE GENUS ROSCOEA
PLATE 13

14. ROSCOEA CAUTLEYOIDES

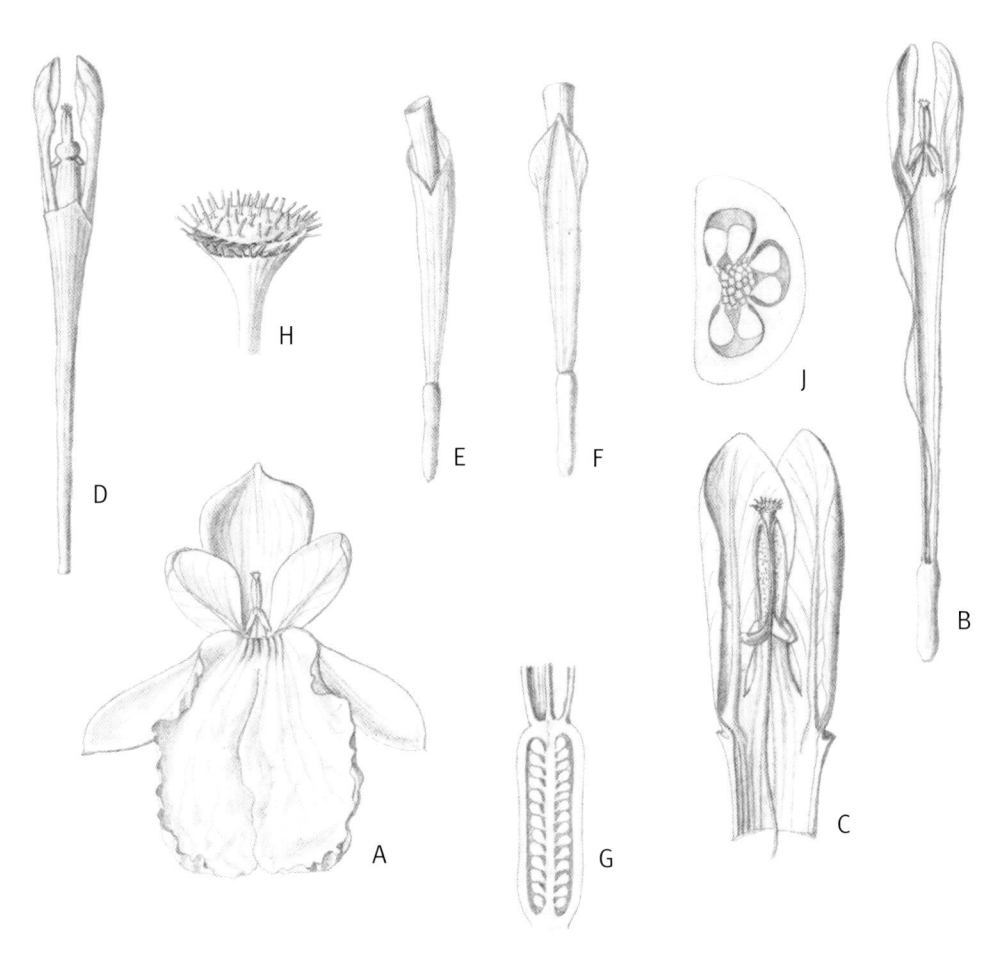

Fig. 54. Dissection diagram of **Roscoea cautleyoides** var. **cautleyoides** forma **cautleyoides** by William Edward Trevithick. **A** flower (front view); **B** staminodes and stamen; **C** cross section showing front view of ovary, style and perianth tube; **D** back view of ovary, style and perianth tube; **E** calyx and ovary (front view); **F** calyx and ovary (back view); **G** longitudinal cross section of ovary; **H** stigma; **J** transverse cross section of ovary.

Another variant of the yellow form is a plant that was informally called *Roscoea cautleyoides* var. *grandiflora* by Mr George Preston, the Assistant Curator in charge of the Rock Garden at Kew, where the plant appeared among self-sown seedlings around clumps of *R. cautleyoides*. He wrote in the *Journal of the Royal Horticultural Society* (Vol. 74 (5): 206 (1950)) that it was "an extremely good form …..with much larger flowers with the same robust habit of the type….". Mr Preston did not formally describe this variant, and it has since been named 'Kew Beauty' by Blooms of Bressingham who distribute plants (Fig. 102). There do not seem to be any records of such a form in the wild. It appears to have been widely distributed. Kew exhibited it at an R.H.S Show in 1957 and it was given an Award of Merit. It is robust and vigorous, and flowers a little later than the typical form.

Plate 13. Roscoea **cautleyoides** var. **cautleyoides** forma **cautleyoides**. *Botanical Magazine* plate 9084, April 1924. Painted by William Edward Trevithick.

14. ROSCOEA CAUTLEYOIDES

Fig. 55 (left). ***Roscoea cautleyoides*** var. ***cautleyoides*** forma ***sinopurpurea*** from Yunnan province, China, CLDX 0687, cultivated at R.B.G., Kew. Photographed by Andrew McRobb, R.B.G., Kew.

Fig. 56 (right). ***Roscoea cautleyoides*** var. ***cautleyoides***. A bronze form growing near Lijiang, Yunnan province, China. Photographed by David Tennant in June 1990.

It has much leafier stems, large wide leaves, larger flowers on a shorter peduncle and grows to about 40 cm, according to Alan Bloom of Bressingham. There is still a large stand on the Rock Garden at Kew. According to a report by an unknown author in the *Journal of the Royal Horticultural Society* in 1958 (Vol. 83 (3)), it comes "more or less true from seed, the flowers are a pale shade of Chartreuse Green (H.C.C. 663/2 to 663/3)…." A good colour photograph of this form appears on page 122 of the *Bulletin of the Alpine Garden Society* Vol. 54 (2) in 1986.

Bees' 1937 catalogue listed a form called 'August Beauty', with the description: "Flowers two months later, vigorous, otherwise like the type. Each 1/-, 3 for 2/4d." Is this still cultivated under this name, I wonder? It was still being mentioned in the literature in the 1950s (e.g., by W.G. Mackenzie in the *Journal of the Royal Horticultural Society* 75 (6): 229–230 (1950), and by A.T. Johnson from Conway, North Wales, in *Gardeners' Chronicle* 133: 23 (1953)) but I have not found any later mention of it. However, there is a late-flowering form still in cultivation and such plants should probably bear this cultivar name.

In 1902 Gagnepain described the purple form of *Roscoea cautleyoides* as a variety of the Himalayan species *R. capitata*, presumably because they both had pedunculate inflorescences (Fig. 55). Rather later, Loesener, working in Berlin, realised that Gagnepain's taxon could not belong to *R. capitata*. He used the same specimens as the type of his new species *R. yunnanensis*. At the same time he described varieties of his new taxon. In 1926 Otto Stapf wrote the text for the *Botanical Magazine* plate (t. 9084) of the yellow form, and at the same time proposed a new species name for the purple-flowered form, *R. sinopurpurea*, but this name is illegitimate as *R. yunnanensis* had been described previously, making Stapf's name superfluous.

In May 1914, Camillo Schneider, who travelled with Heinrich Handel-Mazzetti, found the purple form in Sichuan along the Yanyuan basin north towards Kalaba. These localities may be the easternmost for the whole genus. The first record of the purple form being in cultivation was in 1948 (*Journal of the Royal Horticultural Society* 73 (12): lxxx (1948)); it was said to have been exhibited by Anne, Lady Brocket, of Ware. However, it may well have been in cultivation earlier; references to *Roscoea capitata* being in cultivation in the 1930s may in fact have been referring to *R. cautleyoides* var. *cautleyoides* forma *sinopurpurea*.

In 1988 Zhu Zheng-yin described a new species from Sichuan naming it *Roscoea pubescens*. Later, an illustration of it appeared in *Flora Sichuanica* 10: 599. In 1997, Wu Te-lin reduced the species to varietal rank as var. *pubescens* of *R. cautleyoides*, and this variety is accepted in the latest *Flora of China* account. It differs from the typical variety in its pubescent leaf sheaths and abaxial leaf surfaces, and by the capsule, which is longer than that of the typical variety.

Some of the confusion surrounding this very widely collected and variable species has already been mentioned, but there is more. Gagnepain simultaneously described *Roscoea cautleyoides* and *R. chamaeleon*. Comparison of the types, however, shows that they are conspecific. The type of *R. chamaeleon* was collected in May when the plant showed neither leaves nor a pedunculate inflorescence. The type of *R. cautleyoides* however, was collected in the same locality in July when the leaves were more fully developed and the peduncle of the inflorescence had become obvious. As the two names were published simultaneously, one of them must be chosen as the correct name.

Fig. 57. **Roscoea cautleyoides** var. **cautleyoides** forma **sinopurpurea** in cultivation. Photographed by Chris Grey-Wilson.

14. ROSCOEA CAUTLEYOIDES

Fig. 58 (left). ***Roscoea cautleyoides*** var. ***cautleyoides*** forma ***sinopurpurea*** growing at Muli, Sichuan province, China. Photographed by Phil Cribb, R.B.G., Kew.

Fig. 59 (right). ***Roscoea cautleyoides*** var. ***cautleyoides*** forma ***cautleyoides*** growing in the Gang ho ba, Yunnan province, China. Photographed by Jill Cowley, R.B.G., Kew.

Since the species has been cultivated for many years under the name *R. cautleyoides*, this was the one chosen. Precociously-flowering plants like the type of *R. chamaeleon* were often confused with *R. humeana*; this is discussed further under that species.

When one looks at a living cultivated population of *Roscoea cautleyoides*, it is easy to see how confusion of identity arose, and why botanists who saw only dried specimens were misled. Young plants have only one or two flowers in a slim, pedunculate, tight inflorescence (Figs. 59, 61) and have either no leaves, or very reduced ones, at flowering time. Older well established plants are much taller and more robust, with strong, many-flowered, pedunculate inflorescences and 3 to 5 longer leaves on each plant (Figs. 57, 63).

This species is widely acknowledged as being one of the most attractive and easily cultivated members of the genus. For years the horticultural literature has sung the praises of this "aristocrat with primrose-yellow orchid-like blooms". Clear marking and labelling within the garden is essential, because the plants make a very tardy appearance when every other plant has appeared above

Plate 14. ***Roscoea cautleyoides*** var. ***cautleyoides*** forma ***sinopurpurea***. CLDX 0687. Painted by Christabel King.

THE GENUS ROSCOEA
PLATE 14

ground. However, as soon as they have deigned to make an appearance, they grow rapidly and very soon produce their charming flowers in late May to early June (Figs. 64, 65). With some protection they can be encouraged to flower as early as April, along with the yellow form of *Roscoea humeana*.

The original habitat of this species is in the Chinese provinces of Yunnan and Sichuan. Here it can be found growing in association with grasses, *Rhododendron*, pine and oak trees, at edges of forests, in open stony meadows or shady situations between shrubs and trees or on rocky slopes, at altitudes between 2000 and 3350 m (Fig. 66). In the wild it flowers from mid-May until August.

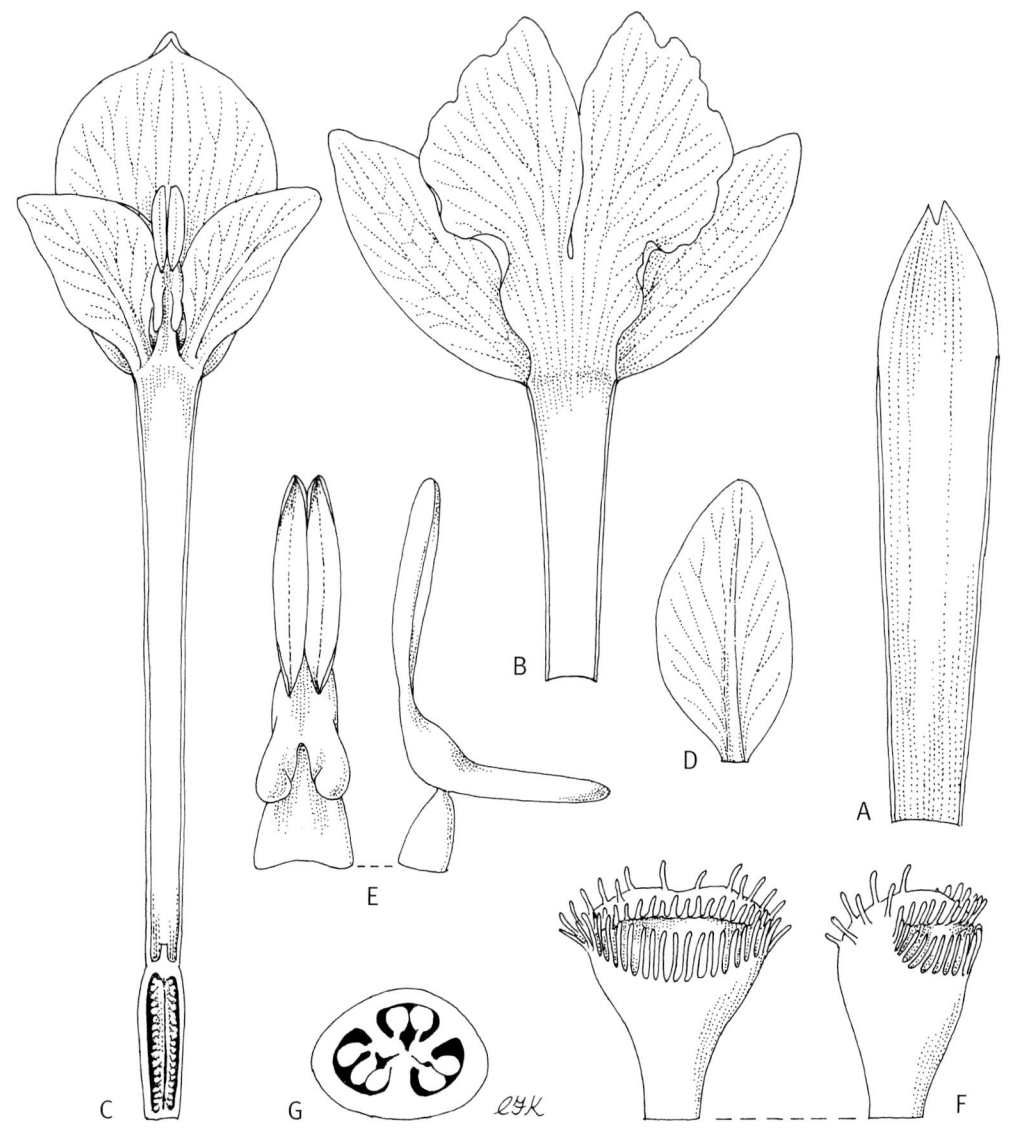

Fig. 60. Dissection diagram of **Roscoea cautleyoides** var. **cautleyoides** forma **sinopurpurea** by Christabel King. **A** inner bract; **B** lower part of corolla; **C** upper part of corolla and ovary; **D** inner perianth segment; **E** stamens, 2 views; **F** stigma, 2 views; **G** ovary, transverse section.

14. ROSCOEA CAUTLEYOIDES

Roscoea cautleyoides Gagnep. in Bull. Soc. Bot. France 48: lxxv (1902). Types: China, Yunnan, Hee-chan-men, 11 July 1883, *Delavay* 231 (syntype P); China, Yunnan, Hee-chan-men, below Lankong near Dali, 11 July 1883, *Delavay* 92 (syntype P, tracing K); near the Erhai Lake, Dali, 19 June, *Prince Henry d'Orléans* (not seen ?P).

R. chamaeleon Gagnep. op. cit.: lxxvi (1902); Hara in Hara *et al.*, Enum. Fl. Pl. Nepal 1: 61 (1978) partly, excl. *Gardner* 525 & *McCosh* 65. Type: China, Yunnan, Hee-chan-men, 21 May 1887, *Delavay* 2659 (holotype P; isotype K).

R. capitata var. *purpurata* Gagnep., op. cit.: lxxiv (1902). Type: China, Yunnan, San-tchang-kiou, July 1889, *Delavay* 4491 (holotype P; isotype K).

R. yunnanensis Loes. in Notizbl. Bot. Gart. Berlin-Dahlem 8: 599 (1923). Type as for *R. capitata* var. *purpurata* Gagnep.

R. yunnanensis var. *purpurata* (Gagnep.) Loes. in Notizbl. Bot. Gart. Berlin-Dahlem 8: 600 (1923).

R. sinopurpurea Stapf in Bot. Mag. sub t. 9084 (1926), nom. illegit. superfl. pro *R. yunnanensis* Loes.

ILLUSTRATION. *Botanical Magazine* plate (t. 9084).

DESCRIPTION. *Plants* (11–)18–35(–55) cm tall. *Sheathing leaves* 3–4, obtuse with pinkish brown markings at veins, sometimes red at base, sometimes densely pubescent. *Leaves* 1–5, or plants sometimes flowering precociously, sometimes glaucous, linear to elliptic, narrowed at the base, widest at the middle, acute to obtuse, scabrid, strongly keeled, upper surface scaley-pustulate, more regularly and densely dotted on lower surface or densely pubescent abaxially, hyaline-edged, side

Fig. 61 (left). ***Roscoea cautleyoides*** var. ***cautleyoides*** forma ***sinopurpurea*** growing in Yunnan province, China, June 1992. Photographed by Jill Cowley, R.B.G., Kew.

Fig. 62 (right). ***Roscoea cautleyoides*** var. ***cautleyoides*** forma ***cautleyoides*** growing in NW Yunnan, China. Photographed by Chris Grey-Wilson.

veins parallel, (7–)15–24(–42) × (1–)2–2.8(–3.2) cm. *Ligule* area triangular in developed leaves, ligule raised 0.5–5 mm. *Inflorescence* shortly to well-exserted from leaves on ridged glabrous peduncle, flowers purple, yellow, white, occasionally pink or rust-coloured, one to several opening at the same time. *Bracts* green with brownish veins, shorter than the calyx, acute, ciliate, first bract large, tubular, enclosing the inflorescence, soon splitting, further bracts decreasing in size, first flower, calyx and subtending bract below main inflorescence, separated by short length of peduncle, sterile bracts sometimes also present at intervals along peduncle, 4–6.3 × 0.9–2 cm. *Calyx* acute, 2.5–5.6 cm long, sharply and deeply bidentate, apiculate, ciliate, occasionally tridentate, veins brownish. *Corolla tube* 3–5.5 cm long, hardly exserted from calyx. *Dorsal petal* obcordate to obovate, 2–3.8 × 1.1–2.7 cm, apex truncate and apiculate. *Lateral petals* oblong to elliptic, 2.4–3.5 × 0.5–1.3 cm. *Labellum* obovate or angular obovate or very broadly truncate obovate, 2.5–4 × 2.3–3.5 cm including claw, deflexed, usually lobed for $1/4$ or more, occasionally entire or trilobed with each lobe emarginate. *Staminodes* asymmetrically obovate or rhombic, 1.2–1.8 × 0.6–0.9 cm, with vein excentric, shortly clawed, lined reddish purple at base. *Anther* pale lemon-yellow, angled below the thecae, thecae 5–6 × 1.5–2 mm, connective elongation in line with the obtuse appendages, together 4–6 × 2 mm, joint between filament insertion and appendages swollen. *Epigynous glands* 4–5 mm long. *Style and stigma* white or cream. *Capsule* 1.5–3.5 × 1 cm. *Seeds* elliptic to triangular, aril deeply lacerate, some lacerations longer than the seed.

One variety is recognised in addition to the typical form.

Fig. 63 (left). **Roscoea cautleyoides** var. **cautleyoides** forma **cautleyoides** cultivated at the R.B.G. Edinburgh. Photographed by Chris Grey-Wilson.

Fig. 64 (right). **Roscoea cautleyoides** var. **cautleyoides** forma **cautleyoides**. Cultivated and photographed by John Fielding, Sheen.

Roscoea cautleyoides var. **pubescens** (Z.Y. Zhu) T.L. Wu in Novon 7(4): 441 (1997). Type: China, Sichuan, Xichang, 2000 m, *J.L. Zhang* 195 (holotype EMA).
R. pubescens Z.Y. Zhu in Acta Phytotax. Sin. 26(4): 315–316 (1988).

DESCRIPTION. Differing from the typical form by having densely pubescent leaf sheaths and lower leaf surfaces and by having longer capsules. I have not seen any material of this variety.

Two colour forms of the typical variety can be recognised and are described here.

Roscoea cautleyoides Gagnep. var. **cautleyoides** forma **sinopurpurea** (Stapf) Cowley, forma nov. a forma *cautleyoide* floribus purpureis nec luteis differt. Typus: China, Yunnan, Wen Bie Shan, near Lijiang, 2800 m, 3 October 1990, *Chungtien-Lijiang-Dali Expedition* CLDX 0687 (holotypus K).

DESCRIPTION. Differing from the typical form (forma *cautleyoides*) by having flowers in the purple range rather than the yellow range.

Another cultivated form that deserves formal recognition was sent to the Royal Horticultural Society on June 11 1963 by Dr J.K. Spearing, a *Roscoea* enthusiast. It is a very dark purple form that he called *R. cautleyoides* var. *purpurea* 'Dark Beauty'. It is much darker than the usual wild purple forma *sinopurpurea*, and is here given the name *R. cautleyoides* var. *cautleyoides* forma *atropurpurea*.

Roscoea cautleyoides Gagnep. var. **cautleyoides** forma **atropurpurea** Cowley, forma nov. a forma *cautleyoide* atque forma *sinopurpurea* floribus atropurpureis nec luteis neque purpureis differt. Typus: Cultivated origin, cult. *Dr J.K. Spearing* (holotypus K).

DESCRIPTION. Differing from the typical form (forma *cautleyoides*) and from forma *sinopurpurea* by having flowers in the deep purple range rather than the yellow or pale to mid-purple range.

Fig. 65. *Roscoea cautleyoides* var. *cautleyoides* forma *sinopurpurea*. Cultivated and photographed by John Fielding, Sheen.

Fig. 66. ***Roscoea cautleyoides*** var. ***cautleyoides*** forma ***cautleyoides*** showing its habitat in the Gang ho ba, Yunnan, China, June 1992. Photographed by Jill Cowley, R.B.G., Kew.

15. ROSCOEA HUMEANA

The more commonly cultivated purple form of *Roscoea humeana*, now forma *humeana*, was beautifully illustrated by W.E.Trevithick in the *Botanical Magazine* of 1925 (tab. 9075). This illustration is reproduced here. Visitors to the Lijiang region of Yunnan would see not only the typical purple form (Fig. 77), but also white and bicoloured forms, and all shades in between (Figs. 69, 70). Here the species is abundant, and especially obvious in mid-summer on the open Lijiang Plain growing with *Incarvillea mairei*, or in groups under the scrubby pines. It also thrives in crevices in the rock outcrops which are scattered over the plain. In the wild, plants nearly always flower before the leaves emerge, a habit not often seen in cultivation.

The yellow form of the species (forma *lutea*), occurs sporadically in Yunnan in predominantly purple- or white-flowered populations. However, in Sichuan, from where the plant illustrated here originates, it is the commonest form (Fig. 72). Herbarium specimens collected in Sichuan are nearly all of the yellow form, although some collectors also mention white and violet-red flowered plants growing with the pale-yellow-flowered ones (Figs. 71, 73). From south-east of Yungning, also in Sichuan, George Forrest recorded a deep rose-purple form (no. 21447). He also collected no. 21437 from north-east of Yungning, describing the flowers as deep, bright yellow with an orange centre. This implies that the yellow form is also variable, since the flowers of the cultivated plants depicted here are pale yellow and concolorous (Figs. 75, 76).

Plate 15. ***Roscoea humeana*** forma ***humeana***. *Botanical Magazine* plate 9075, May 1924. Painted by William Edward Trevithick.

THE GENUS ROSCOEA | 119
PLATE 15

15. ROSCOEA HUMEANA

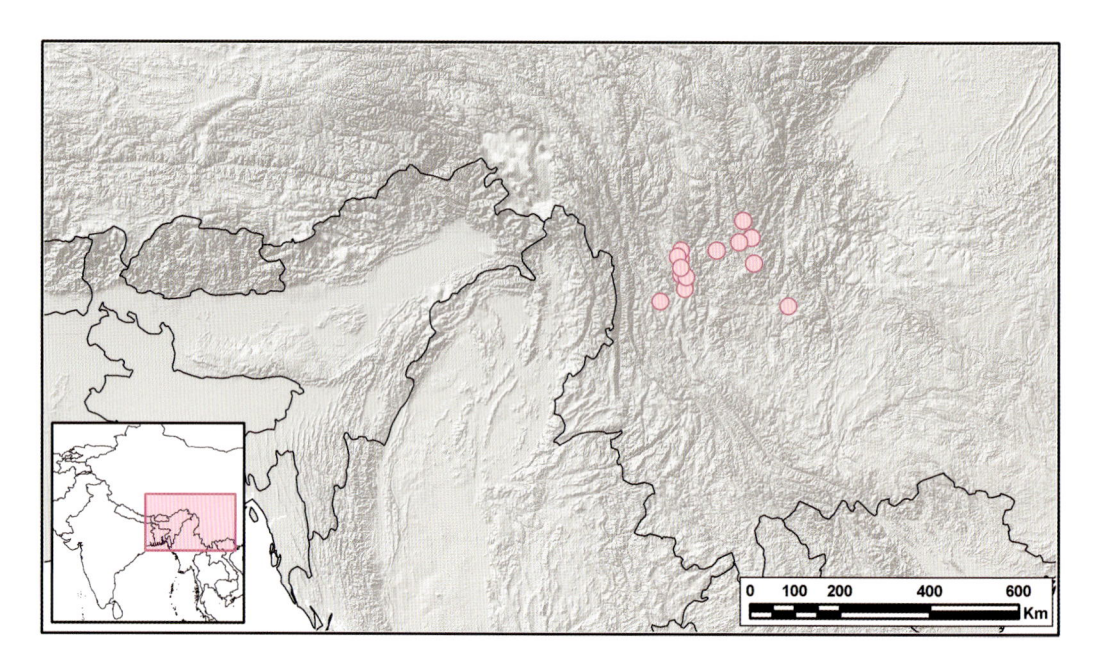

Map 18. Distribution of **Roscoea humeana**.

Fig. 67. **Roscoea humeana** forma **humeana** growing in the Gang ho ba, Yunnan province, China. Photographed by Chris Grey-Wilson.

THE GENUS ROSCOEA
15. ROSCOEA HUMEANA

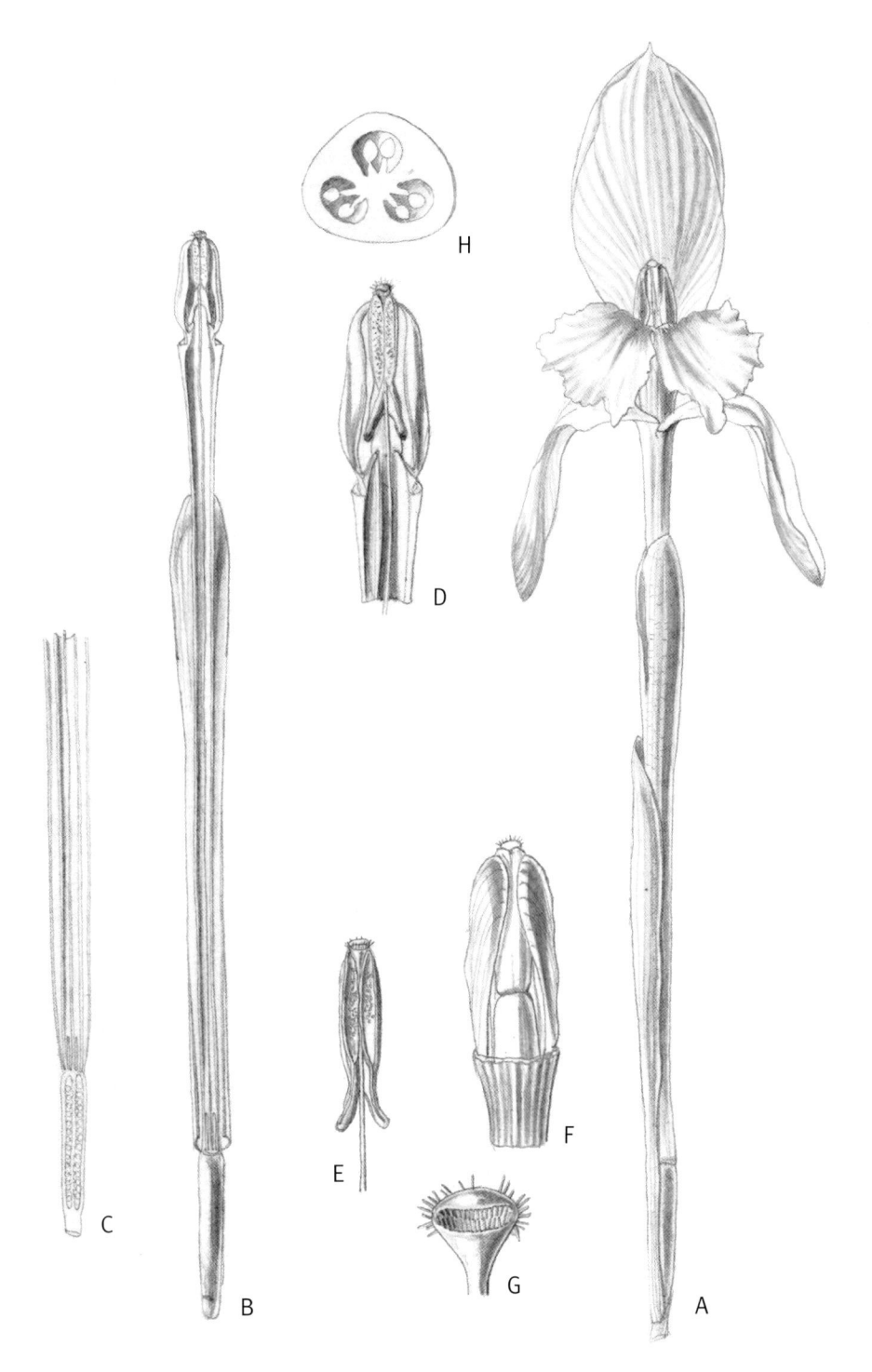

Fig. 68. Dissection diagram of ***Roscoea humeana*** forma ***humeana*** by William Edward Trevithick. **A** flower, bracts and ovary; **B** longitudinal section through flower with petals removed; **C** longitudinal section showing ovary, epigynous glands and calyx; **D** front view of stamen, staminodes and style; **E** stamen, style and stigma; **F** back view of stamen, staminodes and style; **G** stigma; **H** cross section through ovary.

Fig. 69. ***Roscoea humeana*** forma ***humeana*** growing on the Lijiang Plain, Yunnan province, China in June 1992. Photographed by Jill Cowley, R.B.G., Kew.

Fig. 70. ***Roscoea humeana*** forma ***alba*** growing on the Lijiang Plain, Yunnan province, China in June 1992. Photographed by Jill Cowley, R.B.G., Kew.

Fig. 71 (left). ***Roscoea humeana*** forma ***lutea*** and forma ***humeana*** growing at Muli, Sichuan province, China. Photographed by Phil Cribb, R.B.G., Kew.

Fig. 72 (right). ***Roscoea humeana*** forma ***lutea*** growing at Muli, Sichuan province, China. Photographed by Phil Cribb, R.B.G., Kew.

The plant used to illustrate *Roscoea humeana* forma *lutea* was grown at Kew from seeds collected (SICH 1027) at the end of September 1992 during an expedition to Sichuan. The seeds came from wild plants that were up to 20 cm tall when in fruit, growing in Muli County between the towns of Old and New Muli, about 5 km north of the Kangwuliangzhi Pass on Kangwuliangzhi mountain at an altitude of about 3570 m. The plants were growing in the open in a west-facing species-rich alpine meadow, with scattered large trees of *Abies georgei* and *Picea likiangensis*, and regenerating secondary trees and shrubs in an area grazed by yaks. A plant from this collection was shown at the Chelsea Flower Show on 20 May 2002 and received an Award of Merit. There is a report of this in the *Quarterly Bulletin of the Alpine Garden Society* 68(4): 503 (2000).

The plants growing at Kew are thriving and, together with the yellow form of a closely related species, *Roscoea cautleyoides,* are the first of the species to flower, as early as the middle of April. The similarity of the two has led to some confusion that was begun by Gagnepain when he described both *R. cautleyoides* and *R. chamaeleon* in the same paper in 1902. Both were based on specimens (*Delavay* 231 & 2659 respectively) collected at Hee-chan-men in July 1883 and May 1887. Gagnepain may have concluded, from these and other specimens, that there were two species growing in the locality, one with a pedunculate inflorescence and the other with a sessile inflorescence, as well as other differences that he believed to be significant. In fact, he had described two species that were just representatives of a single species (*R. cautleyoides*) at different stages of development. As a result of this, herbarium specimens at an early stage of growth were often identified as *R. chamaeleon*. However, many of these actually belong to the species that we now know as *R. humeana*, which is not found at

15. ROSCOEA HUMEANA

Fig. 73. ***Roscoea humeana*** forma ***lutea*** and forma ***humeana*** with *Incarvillea*, growing on a hillside at Muli, Sichuan province, China. Photographed by Phil Cribb, R.B.G., Kew.

Hee-chan-men, which is near the Erhai Lake, Dali in Yunnan Province, but only in the Lijiang and Yulong mountain areas further north in the same province.

Roscoea humeana was not recognised as distinct and described by Isaac Bayley Balfour and William Wright Smith until 1916. They based their description on plants cultivated at Edinburgh Botanic Garden from seed collected by Forrest, and they compared their new species with Gagnepain's *R. chamaeleon* (now regarded as conspecific with *R. cautleyoides*). They named their new species after Private David Hume, a young gardener at the Royal Botanic Garden, Edinburgh, who fell in action during the Retreat from Mons in 1914.

Balfour and Smith stated that the plants flowered freely during June in 1912–1915, and described them as 'the finest species as yet known in the genus'. The dates suggest that the seeds were collected during Forrest's visit to the Lijiang range in 1910. Studies of herbarium collections of Forrest's specimens suggest that the seeds may have come from his number 5930, collected in June on the eastern flank of the range at an altitude of 3050 m. The plants were found in crevices of dry limestone cliffs.

The species quickly became well-known in cultivation; plants were generously distributed by H.J. Elwes who was growing them in his garden in 1920, possibly from seeds bought from Bees of Chester. After being shown at a Royal Horticultural Society Show in June 1916, it was mentioned under 'other exhibits' at the Floral Committee and then, when shown at the Chelsea Flower Show by Bees in 1920, it was honoured with an Award of Merit.

Plate 16. *Roscoea humeana* forma ***lutea***. SICH 1027. Painted by Christabel King, May 1999.

THE GENUS ROSCOEA
PLATE 16

15. ROSCOEA HUMEANA

As with nearly all species in this genus, there is a considerable amount of infraspecific variation. Cultivated plants of forma *humeana* show clearly that the wide dorsal petal is larger, and certainly longer than the labellum; in other species the latter is usually the dominating perianth segment (Figs. 78, 79). A fine example of this is a collection, now thriving in cultivation at Kew, which was made in 1990 on the joint British-Chinese Chungtien-Lijiang-Dali Expedition under the number CLD 1154A; the labellum is relatively insignificant and the whole flower is dominated by the large dorsal 'bonnet' and the wide lateral petals. However, the yellow-flowered form from Sichuan shown here shows less of a contrast; the labellum is only slightly smaller than the dorsal petal. In other plants it very often equals it

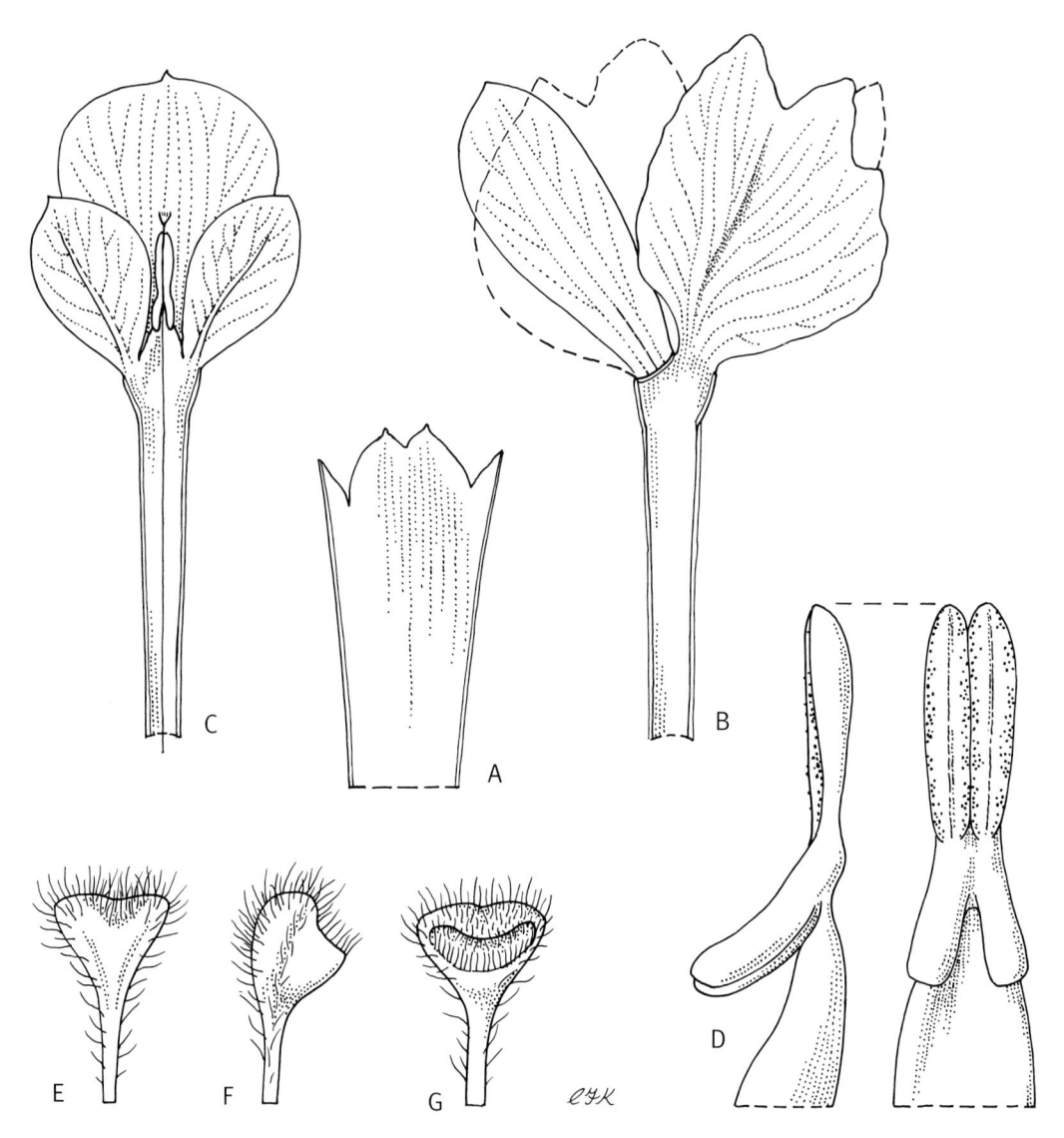

Fig. 74. Dissection diagram of **Roscoea humeana** forma **lutea** by Christabel King. **A** upper portion of floral bract; **B** lower portion of flower with lateral petal and one side of labellum removed; **C** upper portion of flower, dorsal petal and lateral petals with stamen and style; **D** stamen, side and abaxial view; **E** stigma, adaxial view; **F** stigma, side view; **G** stigma, abaxial view.

in size. Purple forms of *Roscoea humeana* have red-coloured shoots on emergence while the yellow forms have green shoots. These variants all share the specific characteristics of *R. humeana*: the very wide leaves; the distinctly exserted bracts and calyces above them; and the calyx that is significantly longer than the supporting floral bract; I have no doubt that they should all be placed in a single species.

By September, most cultivated plants of *Roscoea humeana* are in fruit and the glossy leaves have become extremely large and broad. The capsules are deeply enclosed by the strongly clasping upper leaves (see Nordhagen, 1932). Nordhagen's publication shows a plant grown at Dublin in the fruiting stage and demonstrates the characteristic constriction at the middle of each seed.

Handel-Mazzetti described roscoeas he saw in 1915, on the Lijiang plain, as: 'Growing in large numbers among the bushes was the scarlet-flowered *Roscoea chamaeleon* (*R. humeana* was not described until 1916) the finest of its genus'. This area has been explored extensively and a red flowered form of *R. humeana* has never been recorded. Maybe the translation from the German is at fault, or the meaning is different; there are perhaps similar problems with Delavay's descriptions of the flowers of some species of *Roscoea* as 'rouges'.

Roscoea humeana can be found in south-west China in Yunnan and Sichuan provinces growing on ledges and in crevices of dry limestone cliffs or on grassy, rocky hillsides, screes, alpine meadows, open calcareous pastures, and amongst scrub on margins of thickets and mountain forests. It is found at altitudes between 2900 and 3800 m and flowers from April to July.

Fig. 75 (left). **Roscoea humeana** forma **lutea** cultivated in the Alpine House at Kew. Photographed by Richard Wilford, R.B.G., Kew.

Fig. 76 (right). **Roscoea humeana** forma **lutea** cultivated at Kew. Photographed by Andrew McRobb, R.B.G., Kew (comm. Tony Hall).

15. ROSCOEA HUMEANA

Fig. 77 (left). ***Roscoea humeana*** forma ***humeana*** growing in Lijiang, Yunnan province, China. Photographed by Chris Grey-Wilson.

Fig. 78 (right). ***Roscoea humeana*** forma ***humeana*** cultivated in the Alpine House at Kew. Photographed by Richard Wilford, R.B.G., Kew.

Roscoea humeana Balf. f. & W.W. Sm. in Notes Roy. Bot. Gard. Edinburgh 9, 42: 122–123 (1916). Type: China, Yunnan, cult. R.B.G. Edinburgh from seeds collected by Forrest (E).

ILLUSTRATIONS. A fine photograph of a plant growing in rock crevices appeared on the cover of *The Garden*, volume 118(1) 1993, taken by Christopher Grey-Wilson in the Yulong Shan, N.W. Yunnan.
Quarterly Bulletin of the Alpine Garden Society 68(4): 503 (2000).
The Rock Garden 26(4): fig. 114; text p. 322 (2000).
Nordhagen's *Zur Morphologie und Verbreitungsbiologie der Gattung Roscoea Sm.* on page 23, fig. 6.
Unfortunately, as late as 1964 Valerie Finnis, in the *Journal of the Royal Horticultural Society*, was confusing *Roscoea humeana* with *R. purpurea*. There is a photograph opposite page 158 of volume 89(4) under fig. 53 which shows *R. humeana* but on page 173 she refers to it as *R. purpurea*.

DESCRIPTION. Rhizomatous, tuberous rooted herb, usually flowering precociously in the wild. Sheathing leaves 3–4, obtuse, sometimes red-veined. Leaves 1–2, rarely 3, (up to 6 in cultivation), oblong to ovate, widest at base, 2–30 × 3–6 cm at flowering stage; at maturity acute, ciliate, bright green, white-dotted on upper surface with scaly pustulations, hyaline edged, side veins parallel. Ligule area ± semicircular, raised for 2 mm. Inflorescence many-flowered, without exserted peduncle. Bracts usually broad, lanceolate, pale green to whitish at base, usually much shorter than the calyx, 3–14 × 0.5–1.5 cm, obtuse to truncate, ciliate. Calyx 7–18 cm, narrowly tubular, oblique, obtuse, bilobed, lobes rounded, sometimes apiculate, ciliate. Flowers all shades of purple, yellow or white, one to many opening at the same time. Corolla tube exserted from calyx by 0.5–3 cm, depending on maturity. Dorsal petal obcordate, cucullate, apex truncate and apiculate to obovate and strongly apiculate, 2.5–4.5 × 1.9–3 cm. Lateral petals oblanceolate, 2.8–4.5 × 0.8–1.5 cm. Labellum deflexed, usually smaller than dorsal petal, 2–4.5 × 2.8–3.5 cm, including the 0.8–1.2 claw, obovate, bilobed, usually to more than halfway, sometimes each lobe emarginate. Staminodes usually white tinged with purple, 1.5–1.7 cm, asymmetrically obovate or rhombic, shortly clawed with coloured V-shaped markings at base, vein median to excentric. Filament c. 5 mm. Anther white, angled below the thecae, thecae 6–10 × 2–3 mm, connective elongation in line with the yellow pointed appendages, together 6–8 × 3–4 mm, swollen at the joint with the filament. Epigynous glands 4–5 mm. Style c. 10 mm. Stigma white, turbinate, ciliate. Capsule oblong, c. 2.5 × 0.5 cm. Seeds elliptic, constricted at the middle; aril deeply lacerate, the lacerations not usually longer than the seed.

Roscoea humeana Balf. f. & W.W. Sm. forma **alba** Cowley, forma nov. a forma *humeana* atque forma *lutea* atque forma *tyria* floribus albis nec purpureis neque luteis neque tyreis differt. Typus: China, Lijiang, SBLE 636 (holotypus E).

ILLUSTRATION. A photograph of this form is to be found as plate 4 on page 33 in the *Proceedings of the Seventh International Rock Garden Plant Conference* (Alpines 2001).

DESCRIPTION. Differs from the typical form (forma *humeana*) and from forma *lutea* and from forma *tyria* in its white flowers.

Fig. 79. ***Roscoea humeana*** forma ***humeana*** cultivated at Glen Chantry. Photographed by Chris Grey-Wilson.

This pure white form (SBLE 636) was introduced to cultivation in 1987, growing at the Royal Botanic Garden, Edinburgh. Writing in the 'Alpines 2001' Conference Proceedings (see above), Ron McBeath, who collected the plants, stated that this form 'regrettably does not come true from seed and propagation by division is slow.' This white form is described above as forma *alba*. There is variation even within white-flowered populations; according to Kirkpatrick (1990), the two plants introduced from the Sino-British-Lijiang Expedition in 1987 differed from one another in the colour of the back of the white flowers; one was tinged pink and the other, greeny-yellow.

Roscoea humeana Balf. f. & W.W. Sm. forma **lutea** Cowley in Bot. Mag. 17(1): 27 (2000), Type: China, Sichuan, Muli County, *Fliegner, Howick, McNamara & Staniforth* SICH 1027 (K, holotype). (Fig. 79).

R. *sichuanensis* R.H. Miau in Acta Sci. Nat. Univ. Sunyatseni 34(3): 81–82 (1995). Type: China, Sichuan, *Rock* 23852 (SYS).

The yellow form of *Roscoea humeana* was described as a new species, *R. sichuanensis*, by R.H. Miau in 1995 and a Joseph Rock specimen (no. 23852), collected at Mount Siga in Sichuan, designated as the type. However, in the account of the genus for the *Flora of China* this has quite rightly been placed in synonymy under *R. humeana* by the authors, Wu Te-lin and Kai Larsen.

A good photograph of the plant of this colour form that received the Award of Merit can be seen in *The Alpine Gardener* volume 70: 447 (2002). A description by Tony Hall of the same plant appears in the previously published *Quarterly Bulletin of the Alpine Garden Society*, Vol. 68(4): 503 (2000).

Roscoea humeana Balf. f. & W.W. Sm. forma **tyria** Cowley in Bot. Mag. 17(1): 27–28 (2000). Type: cultivated origin, cult. Mrs C.M. Coller (K, holotype).

There is a very good account of this form by Robert Rolfe in the *Bulletin of the Alpine Garden Society* in December 1998 on pages 473–474.

ILLUSTRATION. A photograph of this splendid plant appeared in the *Bulletin of the Alpine Garden Society,* March 1998: 66.

One clone of this dark form, selected from a batch of variously coloured plants, cultivated by Mrs Kath Dryden, was exhibited by Mrs C.M. Coller at the Alpine Garden Society's Summer Show South in June 1997. It has since been given the cultivar name of 'Inkling', and in June 1998 received a P.C. (Preliminary Commendation) from the Royal Horticultural Society.

Fig. 80. **Roscoea humeana** forma **tyria** (ex Mrs Coller), cultivated at Kew. Photographed by Richard Wilford, R.B.G., Kew.

16. ROSCOEA FORRESTII

This Chinese species could almost be called the poor man's *Roscoea humeana*. It is undoubtedly closely related to that horticulturally more flamboyant species but is more widely distributed in the wild. It is, nonetheless, a very attractive species when several plants grow together and are in full flower. Most plants have yellow flowers, but there is also a purple form, described formally here. The type collection, chosen when this species was first described in 1982, is a Forrest collection from Dali in Yunnan, China, a location where *R. humeana* has not been found.

In the 1930s, McLaren or his collectors found both colour forms growing together on Chi Tsu mountain, the exact locality of which is not certain. He gave the typical yellow form the number B105, and his B106 becomes the type of the purple colour form that is described here. Gebauer found this species between Dali and Yungtschang at 2000 m in July 1914 and Kingdon Ward collected it above the lake at Yungning in Sichuan at 3350 m in May 1921 (*Kingdon Ward* 4104).

The plant depicted here was sent to Kew in 1964 as *Roscoea humeana* var. *lutea* by Perry's Hardy Plant Farm. Another collection of this species that is in cultivation at Kew was made on the Chungtien-Lijiang-Dali Expedition to China in 1990 (CLDX 0774 and 0774A). This collection was made in the Gang Ho Ba to the north of the Lijiang plain in Yunnan on 4 October when flowering was over; it was growing with several other species of *Roscoea*.

Roscoea forrestii resembles *R. humeana* in general habit. The inflorescence arises at the apex of a stem of overlapping leaf-sheaths, below which there are up to three leaves, more or less distichously arranged and widest at the base. Numerous herbarium specimens suggest that *R. humeana* seems to flower more precociously than *R. forrestii*, although the difference seems less well marked in cultivation. The leaves of *R. forrestii* usually seem to be well-developed at flowering time (Figs. 81, 83).

The two species differ in several ways: *Roscoea humeana* has large purple or yellow flowers with large, wide petals longer than the labellum, and relatively small staminodes. *R. forrestii* has smaller

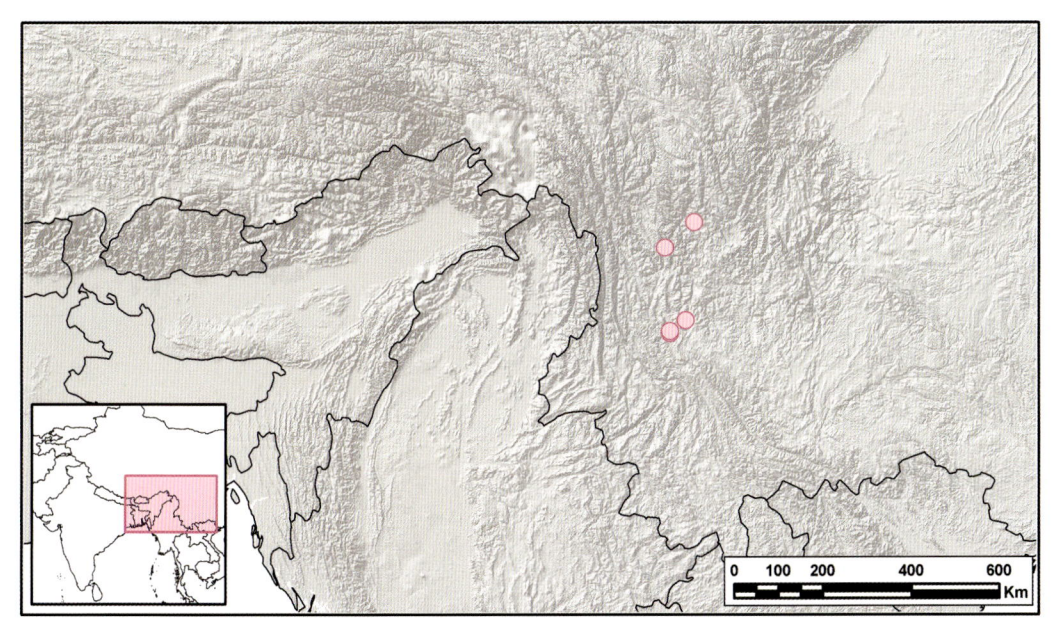

Map 19. Distribution of ***Roscoea forrestii***.

132 THE GENUS ROSCOEA
PLATE 17

flowers with a narrower dorsal petal; the staminodes are larger in proportion to the dorsal petal; the lateral petals are smaller and narrower, and the labellum is equal to or longer than the lateral petals and clearly longer than the dorsal petal. In *R. humeana* the large showy calyx is much longer than its bract; indeed, the bracts are sometimes almost hidden at the base of the inflorescence. In *R. forrestii* the inflorescence and flowers are less showy, and the bracts and calyces are more nearly equal.

Roscoea forrestii can be found in Yunnan and southern Sichuan in crevices and on ledges of cliffs, in open south-facing limestone slopes, and amongst dwarf bamboo and shrubs. It grows up to an altitude of 3350 m, and flowers from the end of May to July.

Roscoea forrestii is named for George Forrest (1873–1932) who made many valuable observations and interesting collections both of living plants and seed (see McLean 2004).

Roscoea forrestii Cowley in Kew Bull. 36 (4): 775–776, Fig. 4 J–Q (1982).
Type: China, Yunnan, Dali range, 3050 m, July 1913, *Forrest* 11726 (holotype K; isotypes BM, E).

DESCRIPTION. *Plants* (17–)20–28(–36) cm tall. *Sheathing leaves* 3–5, obtuse, hyaline-edged, reticulately veined, tinged and dotted pink. *Leaves* 1–3, or plants sometimes flowering precociously, lanceolate to oblong-ovate, acute to obtuse, glabrous or rarely pubescent, mid-green, sometimes scabrid at apex, side veins parallel (6.5–)8–11.5(–13) × (2–)2.7–3.7(–5) cm.

Fig. 81. ***Roscoea forrestii*** forma ***forrestii*** cultivated at Kew. Photograph by R.B.G., Kew.

Plate 17. *Roscoea forrestii* forma *forrestii*. Painted by Christabel King, June 2000.

16. ROSCOEA FORRESTII

Ligule area ± semicircular, ligule raised for 2–3 mm. *Inflorescence* without exserted peduncle, flowers purple or yellow, 1–3 open together. *Bracts* pale green, equal to or shorter than the calyx, sometimes very reduced, sometimes fused together, usually showing just above the leaves, obtuse, scabrid, white-hyaline-edged, 5.2–7.5(–16) × 1–1.5(–2) cm. *Calyx* broad, roundly lobed, bidentate or tridentate, ciliate, tinged pink, 5–13 cm. *Perianth tube* usually well exserted from calyx, 5–12.5 cm long. *Dorsal petal* broadly elliptic with dark veins, apiculate, 2.5–4 × 1.5–2.5 cm. *Lateral petals*

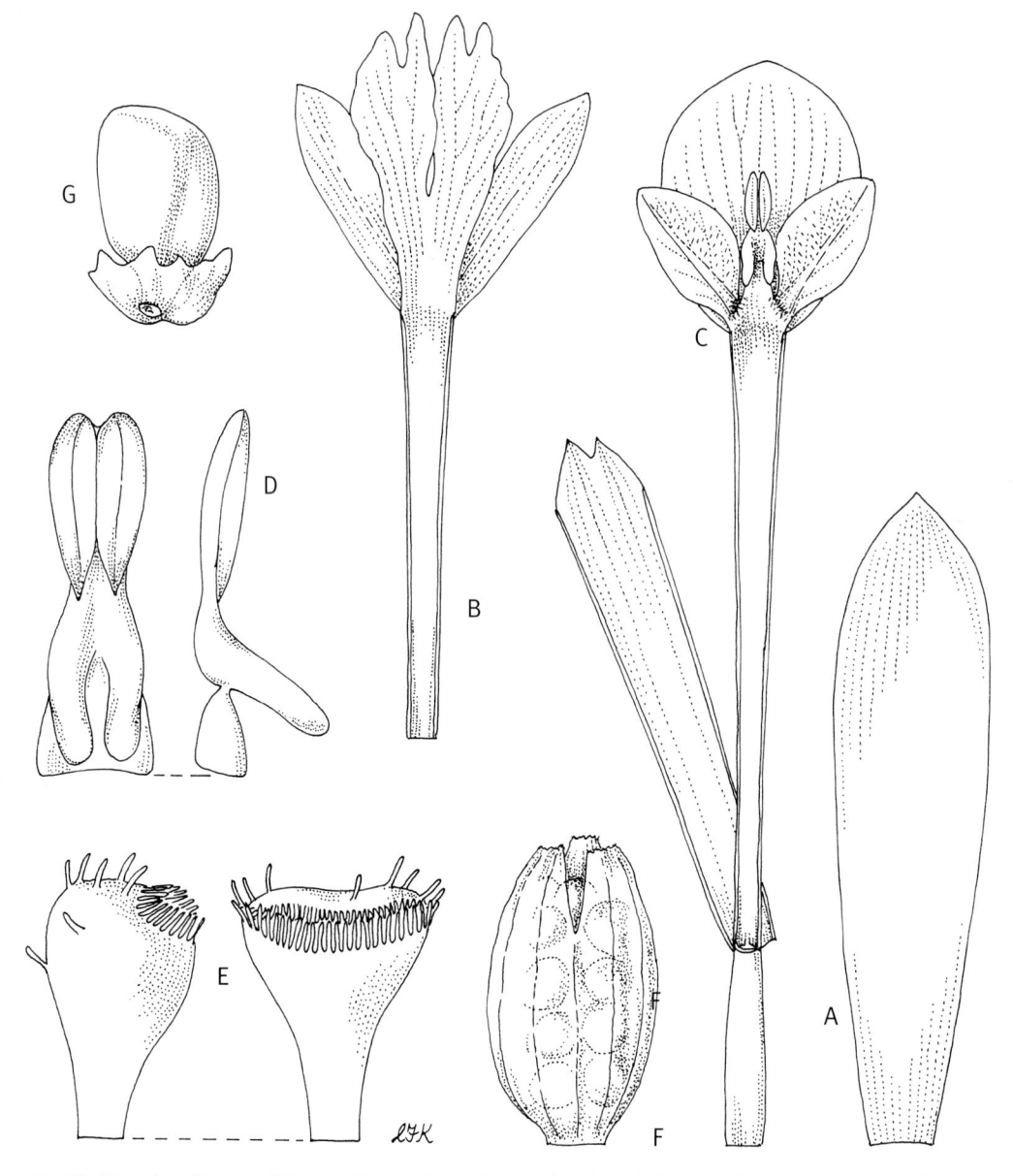

Fig. 82. Dissection diagram of **Roscoea forrestii** forma **forrestii** by Christabel King. **A** bract; **B** lower part of corolla; **C** upper part of corolla and inner bract; **D** stamens, 2 views; **E** stigma, 2 views; **F** capsule; **G** seed.

linear-oblong to elliptic, 2.6–4 × 0.5–1 cm. *Labellum* deflexed, obovate, usually lobed for more than half its length, each lobe emarginate, 3–4.1 × 2.1–3 cm including 6–8 mm claw. *Staminodes* concolorous with petals, sometimes veined white at tip, asymmetrically obovate to rhombic, shortly clawed, vein excentric, occasionally almost median, (1.1–)1.5–2.1 (–2.5) × 0.7–1.1 cm. *Anther* cream, thecae 5–8 × 2 mm, connective elongation angled to the deep bright yellow, obtuse, appendages, together 5–9 × 3 mm. *Ovary* 1–5 cm. *Epigynous glands* 4–5 mm. *Stigma* white. *Seeds* circular or square, aril shallowly lacerate.

Roscoea forrestii Cowley forma **purpurea** Cowley, forma nov. a forma *forrestii* floribus purpureis nec luteis differt. Typus: China, Yunnan, Chi Tsu Mt, June, *McLaren* B106 (holotypus K, isotypi BM, E).

DESCRIPTION. Differs from the typical form (forma *forrestii*) by having purple flowers rather than yellow.

Fig. 83. **Roscoea forrestii** forma *forrestii* cultivated at Kew. Photograph by Richard Wilford, R.B.G., Kew.

17. ROSCOEA TIBETICA

This is one of the most widespread of the Chinese *Roscoea* species in the wild, but is still little known in cultivation. It is certainly not one of the more showy species, but it can tolerate both shaded and open areas and deserves a place in a corner of the rock garden. In the wild, this species, which is widespread in southeastern Tibet, Burma and western China, can be found growing equally happily in the deepest shade of a pine forest or in open meadows.

The plants illustrated, which represent four of the five colour forms of *Roscoea tibetica* described, were found at an altitude of 3480 m in Yunnan, China, in an open marshy meadow near a fast running stream on the Zhongdian (Chungtien) plateau at Beta Hai, at the junction with the road to Haba Shan. These colour forms, and many in between, were found all over the meadow. The plants collected (as ACE 406 A–E) were chosen to represent the colour range within the population, and at the time of collection were no more than 4 cm high. These plants were found during the Alpine Garden Society's Expedition to China in 1994. Other plants now in cultivation at Kew originate from collections made on the Chungtien-Lijiang-Dali Expedition in 1990. The collections were CLD 176, 282, 483, 1059, 1209, 1238, 1154B (Fig. 85) and 1543. *R. tibetica* was found in carpets in deep shade under pines at 2400–2900 m in the Cangshan mountains above Lake Erhai and in several other localities. (Figs. 86, 87).

THE GENUS ROSCOEA
PLATE 18

The ACE collections now in cultivation are forms that are pure white, white with a purple labellum, deep purple, pinkish-purple and deep pink. These forms are described below and given formal names. However, among the twelve other collections of this species growing at Kew, no two are the same; all are variations on a theme of one or other of these colour forms. It seems a nonsense that they should each merit a cultivar name as soon as they come into cultivation, so these other collections will not be distinguished at this stage.

This species is very variable not only in its colour forms (as described above) but also in its habitat and morphology. The leaf shape can vary from almost rounded to narrowly elliptic. An example of the latter state is a collection from western Yunnan, on the Sino-British expedition to China in 1981 (SBEC 1055) which flowers early and has long yellow-green leaves and pale mauve-white flowers. By contrast, CLDX 1154B is a very good, strong form collected on the Chungtien-Lijiang-Dali Expedition in 1990, that has roundish leaves and mid-purple flowers with a white tip to the dorsal petal (Fig. 85). Other collections made during the 1994 ACE expedition, from the slopes of Haba Shan at 3000 and 3160 m respectively, were larger in all their parts and up to 15 cm tall (ACE 346 & 353). They were found growing in shade in moist woodland, and in partial shade under small shrubs. The species occurs over a wide altitudinal range and some of the variation may be correlated with altitude.

The considerable variability of this species means that several varieties that have been described can no longer be upheld. Some were described as varieties of *Roscoea tibetica*; others, as varieties of *R. intermedia* (= *R. alpina*). I have seen the types of these names, and consider them to be synonyms of *R. tibetica*. A full listing is given below.

The first introduction of this species to cultivation appears to be by the Royal Botanic Garden, Edinburgh in 1981, probably from the SBEC expedition (Lancaster 1989). The plants were collected at Huadianba above Lake Erhai in Yunnan. The previous year, *Roscoea tibetica* was mentioned as part of the flora of the Gang Ho Ba, at the end of the Lijiang Plain (Grey-Wilson 1988).

Some plants sold as *Roscoea tibetica* are, in fact, *R. australis* (see p. 86). These plants originated from a population collected by Kingdon Ward (KW 22124) in 1956 on Mt Victoria in Chin State in Burma. The most obvious difference between the two species is that *R. australis* has its leaves arranged distichously and *R. tibetica* has them arranged in a rosette. There are also various differences in floral structures. In *R. tibetica* the labellum is up to 1.8 cm wide and deeply lobed, and the lobes and lateral petals are usually splayed outwards. In *R. australis*, the labellum is up to 2.4 cm wide and shallowly lobed, and the lobes and lateral petals are not splayed outwards.

Roscoea tibetica has sometimes been confused with *R. alpina*, which looks very different at maturity, because the flowering stem elongates and carries the seed capsules, within the leaf sheaths, above the ground. *R. tibetica*, on the other hand, remains relatively squat, and the capsules are to be found at ground level; the leaves spread apart at the base as the capsules mature and split to reveal the seeds (Fig. 6). *R. alpina* can also be distinguished by its rounded, bonnet-like dorsal petal, rounded staminodes and very small floral bract. The pink form of *R. scillifolia* is commonly misidentified as 'alpina' in the nursery trade, and could also be confused with *R. tibetica*. *R. scillifolia* can, however, be distinguished from *R. tibetica* by its inflorescence, which is

---

Plate 18. *Roscoea tibetica* forma **alba** (ACE 406A, lower right), forma **albo-purpurea** (ACE 406B, middle), forma ***roseo-purpurea*** (ACE 406D, lower left) and forma ***rosea*** (ACE 406E, top). Painted by Christabel King, June 1998.

17. ROSCOEA TIBETICA

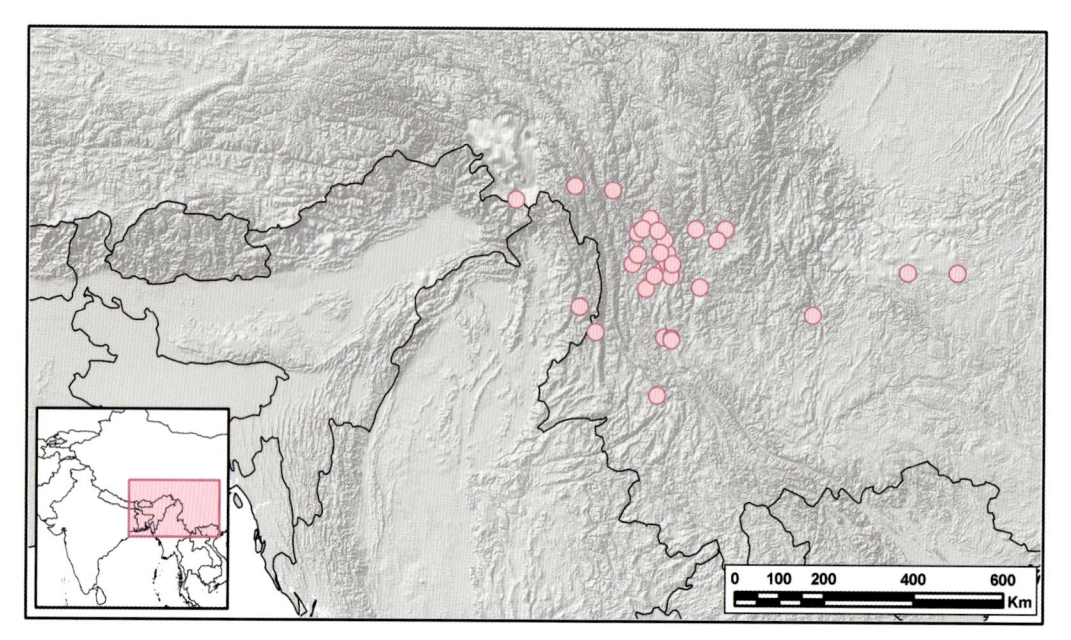

Map 20. Distribution of ***Roscoea tibetica***.

held above the leaves, sometimes on a peduncle, by its smaller flowers and its longer, narrower leaves. The labellum of *R. scillifolia* only attains 2 × 1.2 cm whereas in *R. tibetica* it attains 2.5 × 1.8 cm. The leaves of *R. scillifolia* attain 25 × 2.7 cm, whereas in *R. tibetica* the shorter and broader leaves attain 20 × 5.5 cm.

When I published my revision of the genus in 1982, I believed that the range of distribution of *Roscoea tibetica* extended into southern Tibet and Bhutan. In the ginger family, herbarium specimens cannot always give a full picture. I did, however, note differences between the specimens from Bhutan and Tibet and those from further east. Recently, the plants found in Bhutan and Tibet have been found to be sufficiently distinct for them to be described as a new species, *R. bhutanica* (Ngamriabsakul & Newman 2000). The main differences between the two species lie in the arrangement of the leaves (which is not easy to see in young flowering plants with only one or two leaves), in the calyx and corolla tube lengths, and in the shape of the staminal appendages. In *R. tibetica* the leaves are arranged in a rosette, the calyx is longer than the bract, the corolla tube is long and exserted from the calyx and the anther connective appendages are obtuse. In *R. bhutanica*, the leaf arrangement is distichous, the calyx is equal to or shorter than the bract, the corolla tube is short, usually included within the calyx, and the anther connective appendages are pointed at the tip.

Roscoea tibetica occurs in southeastern Tibet close to Yunnan, in Burma along the Burma-Yunnan borders, and in China in Yunnan and Sichuan Provinces. It grows on open, dry, sunny, grassy or rocky banks, or in moist alpine meadows near rivers or on flat plains. It can also be found in clefts and along cliff ledges on mountain slopes, or in valleys at the margins of, in clearings within, or in the shade of pine and mixed forest or scrub. It has been collected at altitudes between 1800 and 4270 m, and flowers from late May to August.

17. ROSCOEA TIBETICA

Roscoea tibetica Batalin in Trudy Imp. S.-Peterburgsk, Bot. Sada (Acta Hort. Petrop.)14, 8: 183 (1895). Type: China, Sichuan, between Dajian lu and Batang, Litang distr. between Ma-geh-chung and Hokou, June 1893, *Kachkarov* (Iter G. N. Potanini) (holotype LE).

R. intermedia Gagnep. var. *minuta* Gagnep. in Bull. Soc. Bot. France 48: LXXIII (1901 [1902]). Type: China, Yunnan, Maeulchan, July 1889, *Delavay* s.n. (holotype P).

R. intermedia var. *plurifolia* Loes. in Notizbl. Bot. Gart. Berlin-Dahlem 8: 599 (1923). Type: China, Yunnan, in meadows of snowy alpine mountains near Lijiang (Lichiang), 3600 m, July 1914, *Schneider* 2070 (holotype B, presumed destroyed; isotypes G, K).

R. tibetica Batalin var. *emarginata* S.Q. Tong in Bull. Bot. Res., Harbin 12 (3): 247–248 (1992). Type: China, Yunnan, Lijiang, 2900 m, *Tong* 42434 (holotype KUN).

ILLUSTRATION. Lancaster (1989: 271).

DESCRIPTION. *Rhizomatous, tuberous rooted herb* (3–) 7–18 (–30) cm tall; sheathing leaves 3–4, obtuse, cleft. *Leaves* 1–5, ovate to lanceolate, 2.2–20 × 1.3–6 cm, forming a rosette (sometimes first leaf with base narrowed, shortly petiole-like, auriculate), widest towards the base, obscurely to densely hairy especially on young leaves at the acute to obtuse apex; side veins parallel. *Ligule* area ± semicircular, ligule raised 0.5–2 mm. *Inflorescence* enclosed in leaf sheaths, flowers opening above rosette; flowers purple, violet, mauve, pink or white, sometimes 2-toned, one open at a time. *Bracts* included, shorter than calyx, 2.2–4 × 0.4–1 cm, spotted brownish-pink, elliptic, acute, sometimes ciliate.

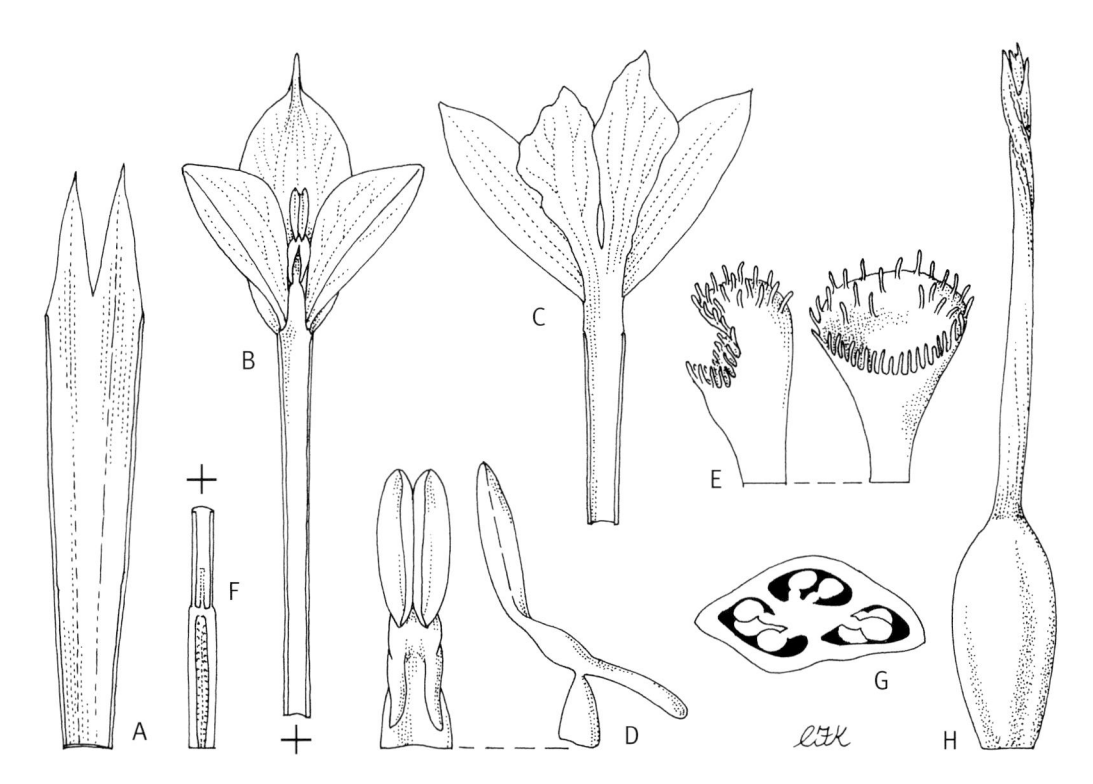

Fig. 84. Dissection diagram of **Roscoea tibetica** by Christabel King. **A** bract; **B** lower part of corolla; **C** upper part of corolla; **D** stamens, 2 views; **E** stigma, 2 views; **F** ovary, longitudinal section; **G** ovary, transverse section; **H** developing capsule.

17. ROSCOEA TIBETICA

Fig. 85. ***Roscoea tibetica***, CLDX 1154B, collected in Yunnan, China and cultivated at Kew. Photographed by Andrew McRobb, R.B.G., Kew.

Calyx exserted, 2.3–3.8 cm, narrow, bluntly, deeply bidentate or tridentate, brown spotted. *Corolla tube* exserted from calyx, sometimes calyx longer than corolla tube, 3–6 cm long. *Dorsal petal* elliptic, hooded, 1.5–2.7 × 0.4–1.5 cm, apiculate. *Lateral petals* usually spreading, linear-oblong, tip acute, 1.3–2.5 × 0.3–0.65 cm. *Labellum* 1.4–2.5 × 0.8–1.8 cm including erect claw, obovate, usually deeply lobed for more than $1/2$ its length, margins sometimes undulate, deflexed, sometimes with white lines at the throat. *Staminodes* elliptic or asymmetrically obovate, 9–18 × 3–6 mm with median vein, sometimes white or with white lines at vein and on edges. *Anther* cream, angled below the thecae; thecae 3.5–7 × 1.5–2 mm; connective elongation in line with the yellow, obtuse appendages, together 3–7 × 2 mm; joint of connective and appendages swollen. *Epigynous glands* 4–5 mm long. *Style and stigma* cream. *Seeds* green to khaki-brown, oblong, round to acute at apex, apiculate, 2–3.5 × 1.5–1.75 mm; arils white, laciniate, as long as seed.

Six colour forms are recognised and described here:

Roscoea tibetica Batalin forma **tibetica**

DESCRIPTION. Flowers white with a pink labellum (holotype LE; see above).

Roscoea tibetica Batalin forma **alba** Cowley, forma nov. a forma *tibetica* floribus omnino albis nec albis labello roseo, vel purpureo, nec omnino atropurpureis, purpureis vel roseis differt. Typus: China, Yunnan, Beta Hai, at the junction with the road to Haba Shan, 18 June? 1994, *Cowley* ACE 406A (holotypus K).

DESCRIPTION. Differs from the typical form in having the flowers pure white rather than white with a pink labellum.

Roscoea tibetica Batalin forma **albo-purpurea** Cowley, forma nov. a forma *tibetica* floribus albis labello purpureo nec roseis labello albo vel omnino albis, atropurpureis, roseo-purpureis vel roseis differt. Typus: China, Yunnan, Beta Hai, at the junction with the road to Haba Shan, 18 June? 1994, *Cowley* ACE 406B (holotypus K).

DESCRIPTION. Differs from the typical form in having flowers that are white with a purple labellum rather than white with a pink labellum.

Roscoea tibetica Batalin forma **atropurpurea** Cowley, forma nov. a forma *tibetica* floribus omnino atropurpureis nec albis labello roseo, albis labello purpureo, vel omnino albis vel roseis differt. Typus: China, Yunnan, Beta Hai, at the junction with the road to Haba Shan, 18 June? 1994, *Cowley* ACE 406C (holotypus K). (Fig. 88).

DESCRIPTION. Differs from the typical form in the flowers, which are deep purple throughout, rather than white with a pink labellum.

Roscoea tibetica Batalin forma **roseo-purpurea** Cowley, forma nov. a forma *tibetica* floribus omnino roseo-purpureis nec omnino albis, atropurpureis vel roseis vel albis labello roseo vel purpureo differt. Typus: China, Yunnan, Beta Hai, at the junction with the road to Haba Shan, 18 June? 1994, *Cowley* ACE 406D (holotypus K).

Fig. 86 (left). ***Roscoea tibetica*** from the Cangshan Mountains, Yunnan, China. Photographed by Jill Cowley, R.B.G., Kew. Fig. 87 (right). ***Roscoea tibetica*** forma **alba** from the Cangshan Mountains, Yunnan, China. Photographed by Jill Cowley, R.B.G., Kew.

17. ROSCOEA TIBETICA

Fig. 88. ***Roscoea tibetica*** forma ***atropurpurea*** (ACE 406C) from the Zhongdian Plateau, Yunnan, China, cultivated at Kew. Photographed by Andrew McRobb, R.B.G., Kew.

DESCRIPTION. Differs from the typical form in the flowers, which are entirely pinkish-purple, rather than white with a pink labellum.

Roscoea tibetica Batalin forma **rosea** Cowley, forma nov. a forma *tibetica* floribus omnino roseis nec omnino albis, atropurpureis vel roseo-purpureis nec albis labello roseo vel purpureo differt. Typus: China, Yunnan, Beta Hai, at the junction with the road to Haba Shan, 18 June 1994, *Cowley* ACE 406E (holotypus K).

DESCRIPTION. Differs from the typical form in the flowers, which are entirely deep pink, rather than white with a pink labellum.

18. ROSCOEA DEBILIS

It is very doubtful that this species has ever been in cultivation. A collection (CLD 1543), made by the Chungtien-Lijiang-Dali (CLD) Expedition on 19 October 1990 and thought to be this species, has now been named as *Roscoea tibetica*. The collection was from the lower slopes of the Cangshan mountain range at Little Huadianba near Dali in the Chinese province of Yunnan. It was thought to be *R. debilis* because of the narrow petiole-like base to the leaves. The plant was way past flowering; if it had been in flower, it would have been obvious that it was not *R. debilis*. Plants distributed by nurseries as *R. debilis* will probably be *R. tibetica*.

Apparently no collections of this species have been made since George Forrest collected it north of Tengchong, in the far west of Yunnan near the Burmese border, in June 1912. From this one could surmise that it may be extinct. However, the distribution was recorded as quite extensive within southern Yunnan and collectors may have concentrated on other areas since then. E.B. Howell found plants in Tengchong in 1911. Forrest made further collections on the eastern flank of the Dali (Cangshan) range, and Augustine Henry collected it in Mengzi in southeastern Yunnan. The type was collected by Liétard for François Ducloux at Lang-ngy-tien, which could possibly be Lang-yen-tsing, near Luteng to the west of Kunming.

Roscoea debilis grows at lower altitudes than most other species, and has not been found above 2440 m. The other species found at low altitudes is *R. praecox*, which is also to be found between Kunming and Mengzi.

Roscoea debilis is quite distinctive. Gagnepain described the plant as frail, graceful and long-stemmed, claiming that its inflorescence resembled that of species in the closely related genus *Cautleya*. The inflorescence is certainly very lax, with the bracts spaced out along the spike, whereas in other species the bracts are contained within the leaves or are clustered at the top of a peduncle. The inflorescence at maturity can be shortly pedunculate or not within the same population, as shown by Forrest's and Henry's collections. Specimens collected from moist situations show flowers exserted from the calyx, with perianth tubes from 3.5 cm long, while those collected in dry places tend to show no exserted perianth tube.

Roscoea debilis resembles *R. cautleyoides* Gagnep. but apart from the lax inflorescence, the staminodes are almost as long as the dorsal petal, whereas in *R. cautleyoides*, they are much shorter. There are two varieties (see below); the specimens of var. *debilis* found near Dali are of the narrow-leaved, smaller-flowered end of the range of specific variation.

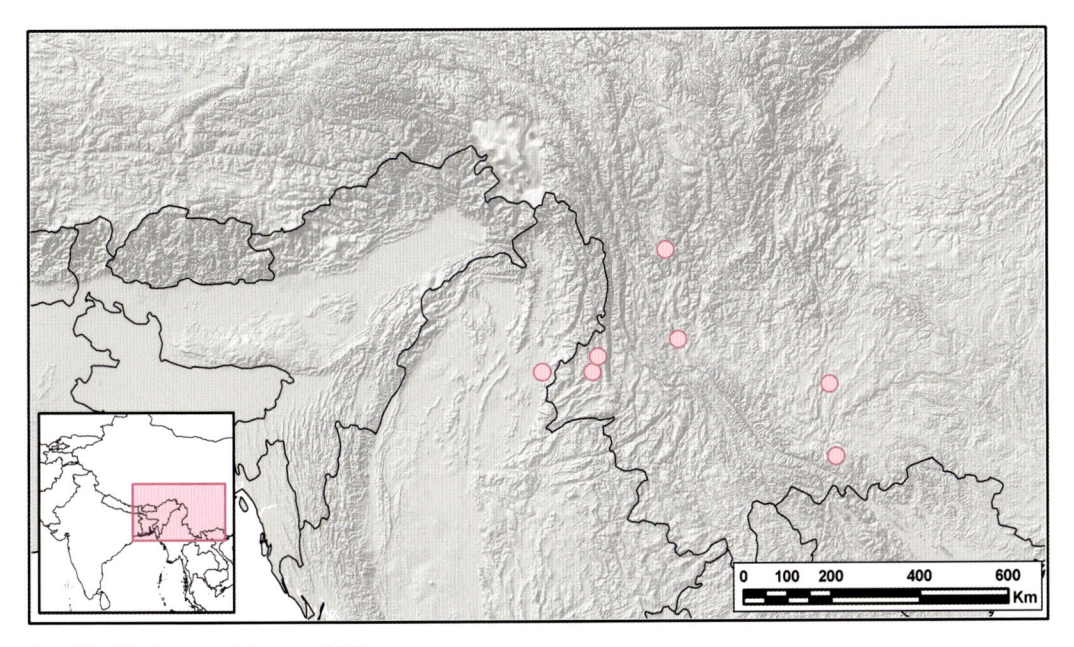

Map 21. Distribution of ***Roscoea debilis***.

The type of *Roscoea blanda* is broad-leaved and has a compact habit (see Fig. 89), while the type of *R. debilis* shows the very lax habit when flowering is nearly over and the leaf-sheaths come away from the main stem. Schumann described *R. blanda* as having sessile leaves, and *R. debilis* as having long flaccid petioles. However, the type of *R. blanda* shows the narrowed, petiole-like bases to the leaves quite well.

Roscoea debilis can be found in grassland, in dry pastures at the base of cliffs, in crevices, and in moist situations on margins of mixed forests at altitudes between 1670 and 2440 m, and flowers from June until August.

Roscoea debilis Gagnep. in Bull. Soc. Bot. France 48: LXXVI (1901).
Type: China, Yunnan, "Lang-ngy-tien" (? = Lang-yen-tsing), Aug. 1899, *Liétard* in *Ducloux* 688 (holotype P).
R. blanda K. Schum. in Engl., Pflanzenr. 4 (46): 121, t. 16B (1904). Type: China, Yunnan, Mengzi, 1830 m, *Henry* 11102c (holotype B, presumed destroyed; isotype K).

DESCRIPTION. *Rhizomatous, tuberous rooted herb*, (13–)18–25(–60) cm tall. *Sheathing leaves* 2–4, obtuse, apiculate, marked pinkish-brown, soon splitting. *Leaves* 2–4, elliptic to lanceolate or oblong-ovate, widest at the middle, acute to acuminate, base narrow and petiole-like, glabrous or pubescent on underside, blade elongating at maturity and parting from the pseudostem by splitting along leaf sheaths, side veins parallel, (5–)8–20(–39) × (1.75–)2–3.5(–5.5) cm. *Ligule* area ± semicircular, ligule raised for 1–2 mm, prominent, pinky-brown. *Inflorescence* sometimes on a shortly exserted peduncle or not, usually only the bracts exserted from the leaves; flowers bluish-purple or white, 1–3 opening together. *Bracts* usually broad, ± equal to the calyx, the first tubular, a short distance below the rest of the inflorescence, obtuse or acute, minutely ciliate, 3.5–6.5 cm × 7.5–15 mm. *Calyx* bidentate, very shallowly notched, obtuse, apiculate, minutely ciliate, spotted. *Perianth tube* exserted or not from the calyx, c. 3.5–7.5 cm. *Dorsal petal* elliptic to narrowly elliptic, strongly apiculate, 2.2–3.8 × 1–1.6 cm. *Lateral petals* linear-oblong to elliptic, 2.2–3.5 × 0.4–0.7 cm. *Labellum* slightly deflexed, narrow to

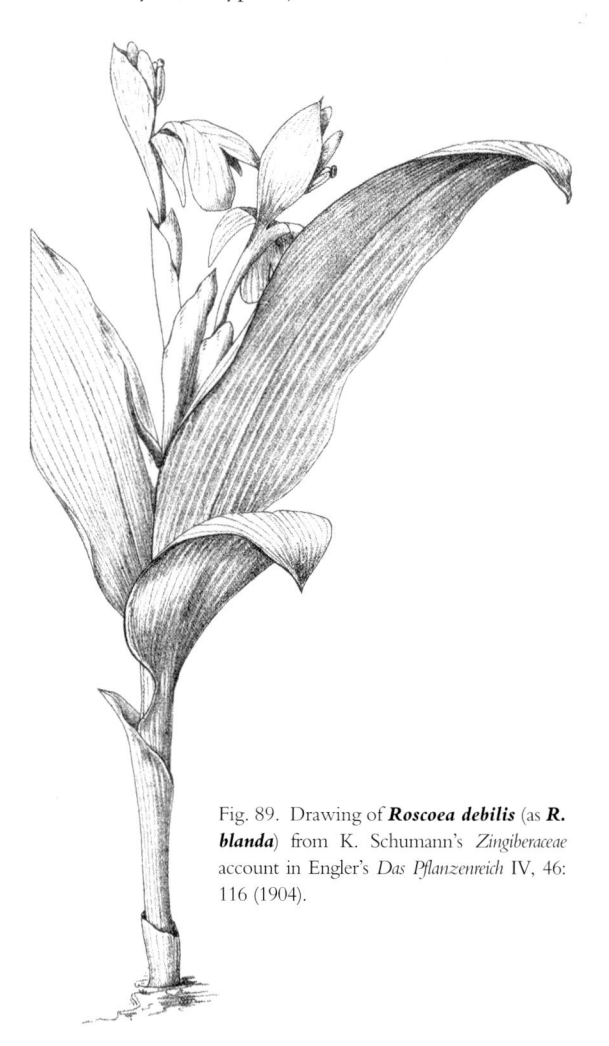

Fig. 89. Drawing of **Roscoea debilis** (as **R. blanda**) from K. Schumann's *Zingiberaceae* account in Engler's *Das Pflanzenreich* IV, 46: 116 (1904).

angular obovate, bilobed, with white lines at throat, 2.3–3.5 × 1.4–5 cm including 5–6 mm claw. *Staminodes* asymmetrically obovate to rhombic, clawed, vein excentric, 1.8–2.5 × 0.6–0.9 cm. *Anther* curved, thecae 6–9 × 1–2 mm, connective elongation usually angled to the pointed appendages, together 7–9 × 2 mm. *Ovary* 1–1.5 cm long. *Seeds* and capsule not seen.

Two varieties can be recognised:
Leaves glabrous beneath . . . . var. **debilis**
Leaves pubescent beneath . . . . var. **limprichtii**

Roscoea debilis Gagnep. var. **debilis**
As type, see above.

Roscoea debilis Gagnep. var. **limprichtii** (Loes.) Cowley in Kew Bull. 36(4): 775 (1982).
R. blanda K. Schum. var. *limprichtii* Loes. in Notizbl. Bot. Gart. Berlin-Dahlem 8: 600 (1923). Type: China, Yunnan, near Dalifu, 2000 m, *Limpricht* 855 (holotype B presumed destroyed; isotype WRSL).

The only material seen of this variety is the type collection from the Dali region. Its main distinguishing feature is the pubescent underside of the leaves; on some plants the pubescence is extremely dense. It is a robust plant, with large oblong-ovate leaves, and the flowers are generally larger than those of var. *debilis*.

19. ROSCOEA PRAECOX

Roscoea praecox appears to have been introduced to cultivation by the Alpine Garden Society's Expedition (ACE) to Yunnan and Sichuan in June 1994. One of the collections (ACE 38) of this species is illustrated here. This and several other species of *Roscoea* collected on this expedition are now growing in the Alpine Department at Kew. (Fig. 91)

The plants were found during the early part of the expedition on 9 June, at the relatively low altitude of 2400 m, on the road north of Dali towards Lijiang. During the Expedition I soon noticed that this plant tended to grow on rocky banks of reddish soil over limestone along the route, and by looking out for this habitat we found the species to be locally common within it. It is an attractive little plant, which occurs in small colonies, but as single plants rather than in clumps. I was also interested to see two different forms, the first to the south and the second to the north of Dali, that appeared recognisably distinct in the field. These differences in habit remain constant in cultivated plants, suggesting that the distinctions seen in the field are not merely induced by local ecological conditions, but have a genetic basis.

The first colony, collected as ACE 19, was discovered on 8 June, 300 km from Kunming between Lanuah and Dali, at 2150 m on a degraded, steep, shrubby hillside under trees of *Pinus yunnanensis*. The impoverished conditions on the disturbed cliff resulted in etiolated *Roscoea* plants up to 15 cm in height. Only one plant showed any vestige of leaf. Some of the inflorescences were held on short peduncles at the top of the stem, with up to four pale or deep purple flowers, opening in succession (Fig. 92). In some plants the prophylls were deep red. Another collection, from a different colony (ACE 38), was made on the following day (Fig. 93). The plant illustrated here came from this second colony. It was growing on flattish areas under depauperate pines, or among

146 THE GENUS ROSCOEA
PLATE 19

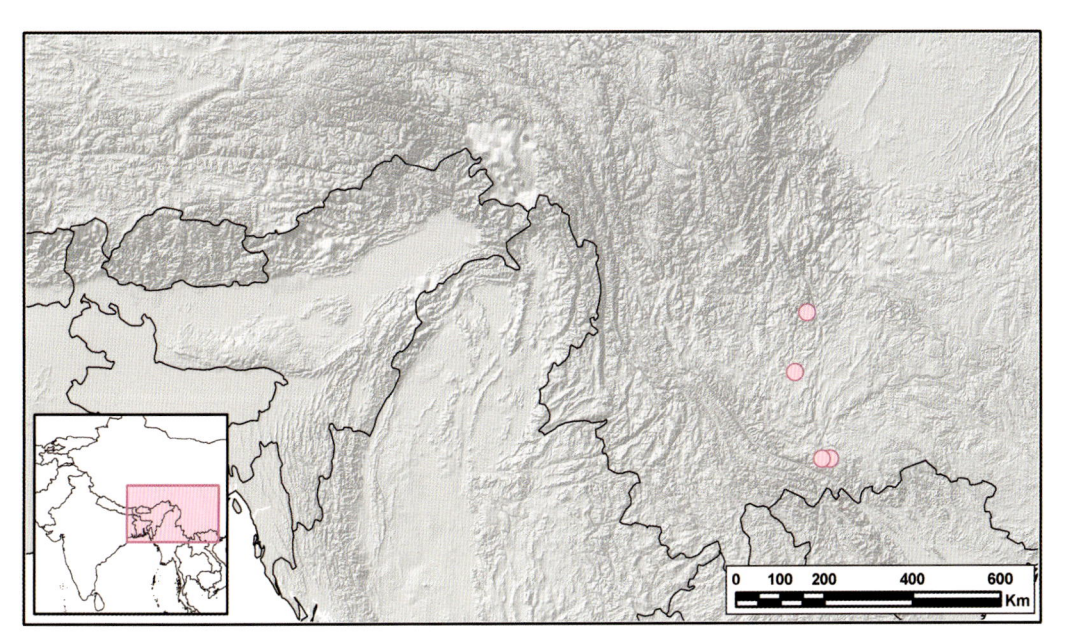

Map 22. Distribution of **Roscoea praecox**.

bushes such as *Hypericum beanii*. A second collection of *R. praecox* at the same site was given the number ACE 38A; they came from a slightly different habitat and were more compact. The inflorescences were not carried on peduncles, so that the effect of the flowers from a distance was of dots of pale to deep purple or pinkish lilac at ground level; only a few had developed very short leaves. A close look at the flowers revealed striking white lines, usually six in number, at the base of the lip (labellum). At this site, *R. praecox* was growing with species such as *Hypoxis aurea*, *Cynoglossum* sp., *Primula poissonii* and a small yellow *Viola*.

Many species of *Roscoea* that are normally leafless at flowering time in the wild tend to flower with the leaves in cultivation. In *R. praecox*, the first collection (ACE 19) has behaved in the same way in cultivation as in the wild; the flowers appear before the leaves. The second collection (ACE 38), on the other hand, flowered with the leaves and is the one illustrated here. Before this expedition, the leaves of this species had not been recorded or described. Schumann (1904), when he described the species, had seen only herbarium specimens, all leafless, and so chose the specific epithet *praecox* which, according to Stearn's *Botanical Latin* (1992), means 'precocious, developing early, over-hasty, bearing flowers before the leaves'. In fact, the leaves can attain a height of almost 40 cm at maturity, but they remain narrowly linear, like those of *R. cautleyoides* and *R. schneideriana* (Cowley, 1994).

Schumann was only able to study a very few specimens when he described this species. More prolific collectors of *Roscoea praecox* such as Schoch, Cavalerie, Gregory and McLaren's Collectors were not in the area until well into the twentieth century. Two specimens, collected by Ducloux at the very end of the nineteenth century were chosen by Gagnepain (1901) as the types of varieties

Plate 19. *Roscoea praecox.* ACE 38. Painted by Christabel King, June 1996.

148 THE GENUS ROSCOEA
19. ROSCOEA PRAECOX

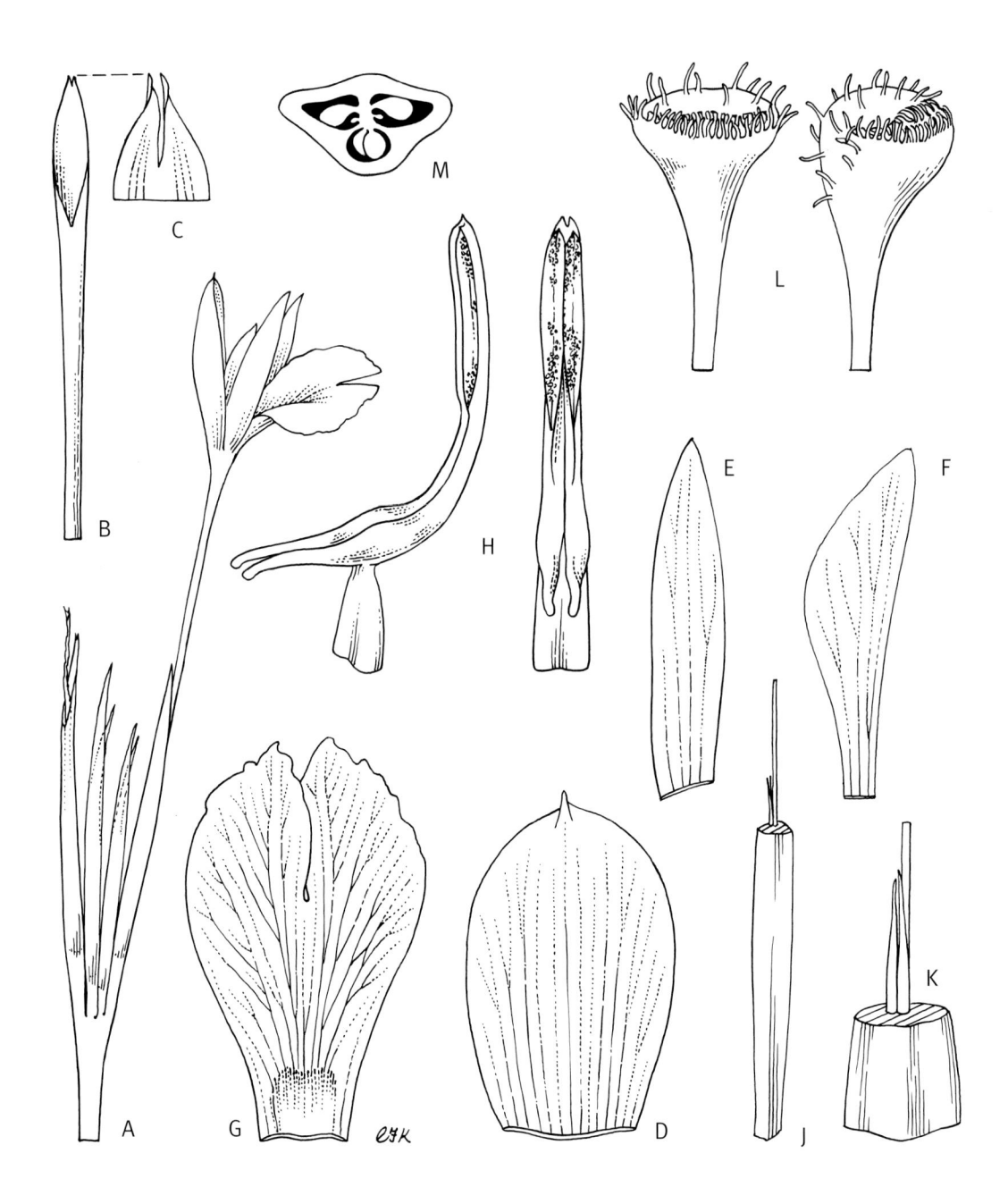

Fig. 90. Dissection diagram of **Roscoea praecox** by Christabel King. **A** inflorescence; **B** bract; **C** detail of apex of bract; **D** dorsal petal; **E** lateral petal; **F** staminode; **G** labellum; **H** stamens, front and side views; **J** ovary, style and epigynous glands; **K** detail of ovary and epigynous glands; **L** apex of stigma, 2 views; **M** transverse section of ovary.

Fig. 91 (left). ***Roscoea praecox***, ACE 38, cultivated at R.B.G., Kew. Photographed by Richard Wilford, R.B.G., Kew.
Fig. 92 (right). ***Roscoea praecox***, NW of Kunming, Yunnan, China. Photographed by Chris Grey-Wilson.

macrorhiza and *anomala* of his *R. intermedia* (a synonym of *R. alpina* from the Himalaya). In my view, both these taxa fall within the range of variation of *R. praecox*, and both names must be placed in the synonymy of that species.

The species whose flowers appear most similar to *Roscoea praecox* is *R. schneideriana*. However, these two species differ in flower colour, stigma and anther appendage shape, as well as in leaf habit, which make them distinct. *R. debilis* is another apparently related but little-known species which also occurs at relatively low altitudes, but it is distinct in certain of its floral characters (Cowley, 1982). *R. debilis* has a very lax inflorescence on an exserted scape with the floral bracts well spaced along it, appearing above the well developed, petiolate leaves. *R. praecox* has an inflorescence which is not exserted, and no, or at best rudimentary, leaves at flowering time (Fig. 93).

Many species of *Roscoea* have been found only in restricted areas, and *R. praecox* falls into this category. It has so far been found only in small colonies in Yunnan, south of Lijiang, on the hills on the outskirts of Kunming, and south of Kunming at Mengzi. Regular visitors to this part of China will know that the landscape is changing rapidly because of logging of mountain forests, damming of rivers, and the destruction of lakeside habitats. Conservation *in situ* is of primary importance but the maintenance of small collections of this potentially endangered species in botanic gardens is an additional safeguard against sudden loss in the wild.

19. ROSCOEA PRAECOX

Fig. 93 (left). ***Roscoea praecox***, Dali to Zhongdian, Yunnan, China. Photographed by Chris Grey-Wilson.

Fig. 94 (right). ***Roscoea praecox***, Lijiang to Dali, Yunnan, China, 23 June 1992. Photographed by Jill Cowley, R.B.G., Kew.

This species is restricted to Yunnan Province in Western China. Plants can be found by banks of streams, in grassy mountain pastures, on shady limestone rocks, dry north-facing banks, pine woodland and shrub-clad hillsides in open positions at 1520 to 2500 m altitude, flowering from the end of April to June (to August in cultivation).

Roscoea praecox K. Schum. in Engl., Pflanzenr. 4 (46): 122 (1904). Type: China, Yunnan, Mengzi, 1525 m, *Henry* 11117 (isotypes E, K).

R. *intermedia* Gagnep. var. *anomala* Gagnep. in Bull. Soc. Bot. France 48: LXXIII (1902). Type: China, Yunnan, Kunming, 18 June 1899, *Ducloux* 597 (holotype P).

R. *intermedia* var. *macrorhiza* Gagnep. loc. cit. (1902). Type: China, Yunnan, Kunming, May 1898, *Ducloux* 601 (holotype P).

ILLUSTRATION. *Bulletin of the Alpine Garden Society* 64: 241, 1996.

DESCRIPTION. *Rhizomatous, tuberous rooted herb 7–28 cm tall. Sheathing leaves* 2–5, herbaceous, obtuse, sometimes deep purple-red. *Leaves* not developed at flowering time, later up to 4, linear, to 40 × 2 cm; leaf sheath undifferentiated at area of join with leaf; *ligule* area triangular to semi-circular, 1–1.5 mm high. *Inflorescence* exserted with peduncle showing or not. *Bracts* greenish, narrowly ovate, widest at base, acute or acuminate, apiculate, 3.7–9 × 0.3–1.2 cm, primary bract tubular at base for 1.2 cm, others free to base. *Ovary* 0.9–1.7 cm long. *Flowers* pale, dark or pinkish purple, up to 4 per inflorescence; *calyx* shorter than to slightly longer than the bracts, tubular, bluntly to sharply bidentate, apiculate, 3–4.8 cm long; *corolla tube* exserted for 0.5–3 cm from calyx, up to 7 cm long; *dorsal petal* elliptic to narrowly elliptic, apex hooded, apiculate, 2.5–3.5 × 0.9–1.8 cm; *lateral petals* linear-oblong, 2.6–3 × 0.4–0.6 cm; *labellum* deflexed, 2.5–4 × 1.6–2.4 cm including 7 mm claw, obovate, bilobed, sometimes each lobe emarginate, with about six strong white markings at junction of limb and claw. *Staminodes* concolorous, white or white with purple at edges, rhombic to falcate, with weak excentric vein, inner edge papillose, apex acute to acuminate, 1.6–2.5 × 0.4–0.8 cm, including narrow claw. *Anther* cream, angled below thecae; thecae 0.55–1 cm; connective elongation in line with the yellow obtuse appendages, together 0.7–1 cm. *Epigynous glands* 3–5 mm. *Stigma* galeate. *Seeds* not seen.

20. ROSCOEA KUNMINGENSIS

I have not been able to see the type specimens of this species and its variety. This account has to rely solely on Tong Shao-quan's original description and a photocopy of the type specimen of the variety.

Plants purporting to be *Roscoea kunmingensis* are already on sale in nurseries (Fig. 95) and it would be interesting to compare them with *R. praecox*. Like the latter, it comes from low altitudes around Kunming and is described as being very much like *R. praecox* but with much smaller flowers, with a more deeply lobed labellum, and small tubular bracts, although this last character is not obvious from the illustration accompanying the description. The illustration compares the dorsal petals, labellums, bracts and staminodes of the two species; the latter differ markedly in shape. The plate shows that *R. kunmingensis* flowers precociously, like *R. praecox*, and also shares the striking white linear marks at the base of the labellum.

Roscoea kunmingensis can be found in pine forests, flowering in May to June at altitudes between 2100 and 2200 m.

Roscoea kunmingensis S.Q. Tong in Bull. Bot. Res., Harbin 12 (3): 248–249 (1992). Type: China, Yunnan, Kunming, 2200 m, 25 May 1990, *S.Q. Tong 42425* (holotype KUN).

DESCRIPTION. *Rhizomatous, tuberous rooted herb* 8–12(–35) cm tall. *Sheathing leaves* 4, membranous, tubular. *Leaves* not developed at flowering time, later narrowly lanceolate to lanceolate, apex shortly acuminate to acuminate, both surfaces glabrous, 8–20 × 2.5–4.2 cm. *Ligule* somewhat semicircular, glabrous, c. 3 mm long. *Inflorescence* not exserted on a peduncle, 1–2 flowered, flowers reddish-purple. *Bracts* tubular, membranous, glabrous, apex acuminate, bidentate, white, sometimes elongating after flowering, 0.5–3.5(–8) cm long. *Calyx* tubular, glabrous, apex bidentate, 2.5–3 cm long. *Corolla* tube 3.5–4 cm long. *Dorsal lobe* oblong, apex mucronate, 1.5–2 × 0.6–0.8 cm. *Lateral lobes* linear-oblong, apex entire, 3–4 mm wide. *Labellum* obovate-cuneate, deflexed, deeply bilobed, narrow, with linear white markings at junction of limb and claw, 1.6–2.1 × 1–1.5 cm. *Staminodes* narrowly obovate-cuneate, nerve excentric, c. 1.4 × 0.8 cm. *Anther* thecae white c. 4 mm long, connective yellow, appendages c. 3 mm long. *Ovary* c. 7 mm long. *Style* linear, white, glabrous. *Stigma* white, ciliate. *Epigynous glands* linear, c. 3 mm long.

20. ROSCOEA KUNMINGENSIS

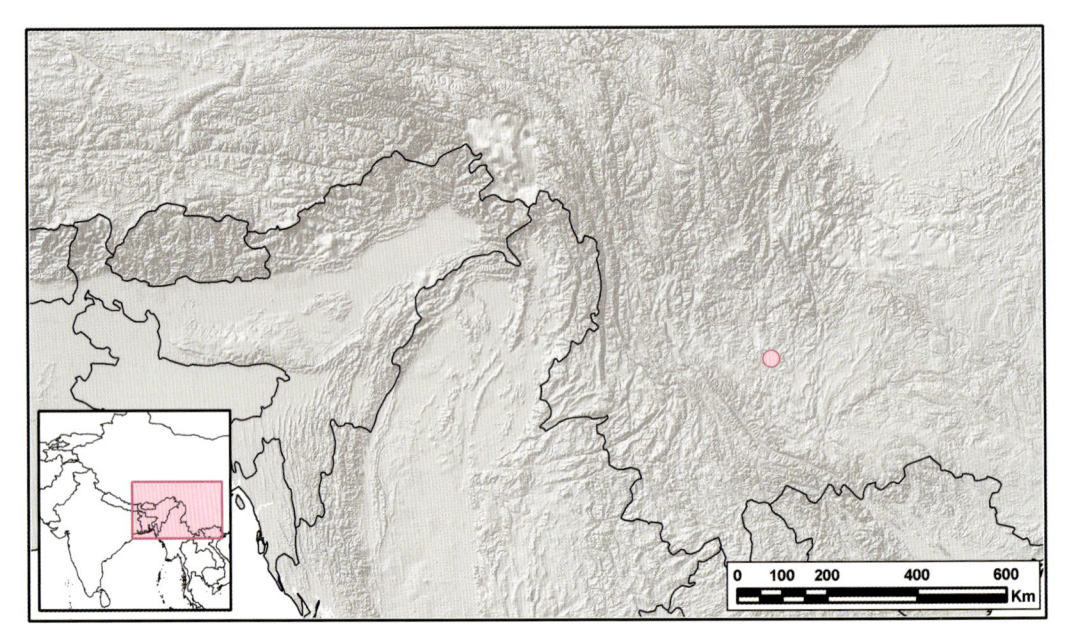

Map 23. Distribution of ***Roscoea kunmingensis***.

Capsule when mature cylindric, yellow-green, calyx persisting, 3–3.5 × 0.6–0.7 mm. *Seeds* obovate, "yellow-green", enclosed by white aril, c. 3.5 mm long.

Two varieties can be distinguished:

Leaves 2.5–3 cm wide; bracts 0.5–0.7 cm . . .
. . . . var. *kunmingensis*
Leaves to 4.2 cm wide; bracts 2.5–3.5 cm, to 8 cm after flowering . . var. *elongatobractea*

Roscoea kunmingensis S.Q. Tong var. **elongatobractea** S.Q. Tong in Bull. Bot. Res., Harbin 12(3): 249–250 (1992). Type: China, Yunnan, Kunming, 2100 m, 21 June 1990, *S.Q. Tong* 42427 (holotype KUN).

DESCRIPTION. Differing from the typical variety (var. *kunmingensis*) by having wider leaves to 4.2 cm and longer bracts, 2.5–3.5 cm long, elongating to 8 cm after flowering. Flowers in June.

Fig. 95. ***Roscoea kunmingensis*** cultivated and photographed by Roland and Gay Bream, Cruckmeole.

9. HYBRIDS AND CULTIVARS

Numerous cultivar names for species of *Roscoea* have been listed by nurserymen in recent years. Many of these are new names for clones that have been known in cultivation for many years. It is almost impossible to sort out which names belong to genuinely new cultivars, and which are superfluous names for old ones. Cultivar names have also often been misapplied and are now attached to the wrong plants, and some have been applied to the wrong species. Sometimes the same cultivar name appears under more than one species.

In this chapter, the names that have appeared in the horticultural literature and in nurserymens' catalogues over the last eighty years or so are listed.

One easily recognisable hybrid is *Roscoea* × *beesiana,* and this plant is discussed here, and illustrated as Plate 20.

Fig. 96 (left). **Roscoea × beesiana** cultivated at Kew. Photographed by Richard Wilford, R.B.G., Kew.
Fig. 97 (right). **Roscoea × beesiana** cultivated and photographed by John Fielding, Sheen.

A list and notes by Roland Bream follows, covering those hybrids and cultivars that are grown by him as the National Council for the Conservation of Plants and Gardens (NCCPG) collection holder for *Roscoea*. This collection is kept at Harnage, Shrewsbury.

Finally a short list is given of suppliers of *Roscoea* species, hybrids and cultivars to the horticultural trade.

PUBLISHED HYBRID AND CULTIVAR NAMES AND THEIR SPECIES

Roscoea auriculata	'Floriade'	
	'White Cap'	
R. auriculata* × *R. cautleyoides	× beesiana	(Figs. 96, 97)
	× beesiana 'Monique' (white fls.)	(Fig. 98)
	× beesiana 'White Form' (creamy white fls.)	
R. cautleyoides	'August Beauty' (late)	
	'Bees' Dwarf'	
	'Beesii' (yellow; late; stout)	
	'Blackthorn Strain'	(Fig. 99)
	'Dark Beauty' (deep purple)	
	'Early Purple'	
	'Grandiflora' (= 'Kew Beauty')	
	'Himalaya' (white)	
	'Jeffrey Thomas' (creamy white)	(Fig. 100)
	'Kew Beauty' (yellow, stout)	(Fig. 103)
	'Purple Giant'	(Fig. 102)
	'Reinier' (yellow)	(Fig. 104)
R. humeana	'Inkling'	(Fig. 80)
	'Rosemoor Plum' (probably = 'Inkling')	
R. purpurea	'Brown Peacock'	(Fig. 105)
	'Nico'	
	'Peacock'	(Fig. 105)
	'Peacock Eye'	(Fig. 105)
	'Red Cap'	
	'Short form'	
	'Tall form'	
	'Vincent'	
R. tumjensis	'Purple King'	

THE GENUS ROSCOEA
HYBRIDS AND CULTIVARS | 155

Fig. 98. ***Roscoea*** × ***beesiana*** and **'Monique'** (left) cultivated and photographed by Roland and Gay Bream, Cruckmeole.

Fig. 99. ***Roscoea cautleyoides*** **'Blackthorn Strain'** cultivated and photographed by Roland and Gay Bream, Cruckmeole.

Fig. 100. *Roscoea cautleyoides* **'Jeffrey Thomas'** cultivated at Cruckmeole and photographed by Richard Wilford, R.B.G., Kew.

21. ROSCOEA ×BEESIANA

It has not been possible to match plants sold under the name of *Roscoea 'beesiana'* with wild collected material. In general plants with the epithet *beesii*, *beesiana*, or *bulleyana* emanated from Bulley's nursery, Bees Ltd. Its owner, A.K. Bulley, sponsored plant collectors such as Forrest, Farrer, Kingdon Ward and Cooper in the Sino-Himalayas to collect plants for sale in his nursery (McLean 1997). He was very keen that un-named plants should be given specific epithets which would draw attention to his firm.

However, the plant collectors whom he sponsored were active in the first half of the twentieth century, but the name *Roscoea beesiana* did not appear until 1970, in Vol. 167 of the *Gardener's Chronicle*. The editor of the Chronicle replied to a letter he had received from Norman Hart of Uckfield, Sussex, which stated: "…….*R. beesiana*, a very robust plant growing to 15 inches or more with truly magnificent flowers of pale yellow streaked with purple. Quite a rare plant in cultivation, it is well worth seeking in specialist's catalogues". In reply the Editor noted: "*R. beesiana* is not a species but a variety of *R. cautleyoides* and is therefore correctly *R. cautleyoides beesiana*".

The next mention of this plant was in 1981 (*Journal of the Royal Horticultural Society* Vol. 106) in an article entitled *Weedy Natured Plants* by Alan Bloom who wrote: "Roscoeas are not commonly grown..… though *R. beesiana* …. causes no problems". In 1987 Roy Lancaster, in his *Garden Plants*

Plate 20. *Roscoea* ×*beesiana* cultivated at R.B.G., Kew. Painted by Christabel King, July 1999.

THE GENUS ROSCOEA | 157
PLATE 20

for Connoisseurs, has a photograph (pp. 160–161) showing plants with very varying degrees of purple on the labellum. The photograph was taken at Bressingham Gardens in Norfolk in July.

Roscoea × *beesiana* is undoubtedly a hybrid, and the unrecorded cross was probably made at Bulley's nursery, Bees Ltd. It is interesting that it is a hybrid between a Chinese species and an Himalayan one. The plant shows the basic colour and small staminodes of the Chinese *R. cautleyoides*, and the purple streaks, large exserted bracts and auriculate leaves of the Himalayan *R. auriculata*. The cross would have to have been made at the end of the flowering period of the former

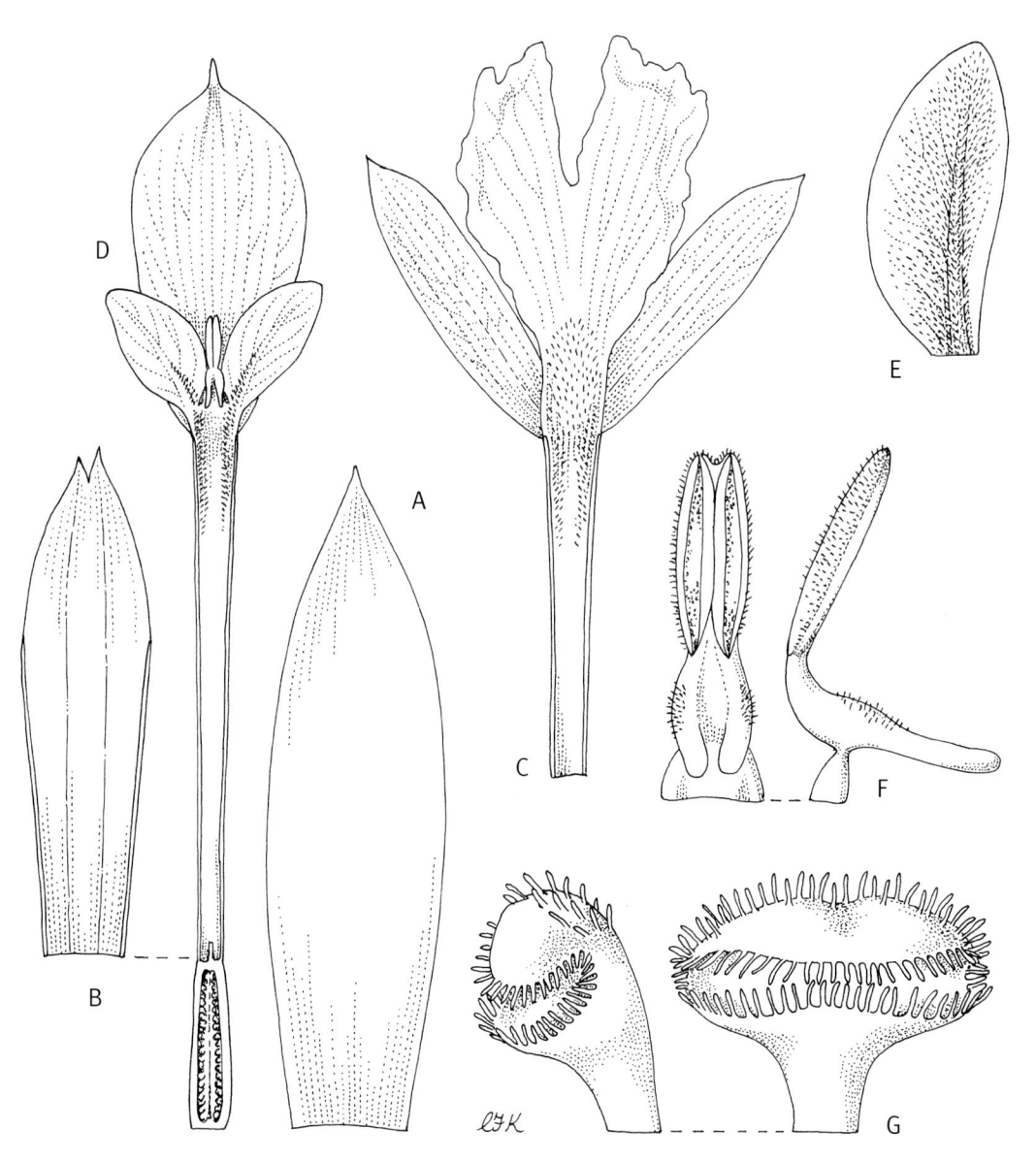

Fig 101. Dissection diagram of **Roscoea × beesiana** by Christabel King. **A** bract; **B** bract (inner); **C** lower part of corolla; **D** upper part of corolla and ovary; **E** inner perianth segment; **F** stamens, 2 views; **G** stigma, 2 views.

and towards the beginning of the latter. In cultivation *R. cautleyoides* is one of the first of the species to flower, while *R. auriculata* waits until later in the year, although there are late-flowering forms of *R. cautleyoides*, and sometimes these flower for a second time later in the year.

The plant illustrated here was donated to Kew by the Dutch firm of Dix and Zijerveld in 1993. Various forms of *Roscoea × beesiana* are offered by nurseries at present: 'Monique' has almost white flowers; 'White form' has creamy-white flowers, and 'Gestreept' has fairly pronounced streaking on the labellum.

Hybrids and Cultivars of *Roscoea* held by the National Collection holder, Roland Bream.
Colour codes are taken from the RHS Colour Chart (2001 Edition).

Roscoea auriculata

'Pink form'. A paler shade of purple (N78B — purple group) compared with the species (77A — purple group).

'Floriade'. Slightly larger flowers than the typical form, and somewhat paler (N81C — purple/violet group). Given an award in 1998 by the KAVB (Royal Dutch Bulbgrowers).

Roscoea × beesiana (see above; probably *R. cautleyoides × R. auriculata*) (Figs. 96, 97). This has never been found in the wild. The flowers are pale yellow (1D — green/yellow group) with variable amounts of purple streaking on the labellum.

'Beesiana Gestreept'. Flowers with fairly pronounced streaking on the labellum.

'Beesiana Monique'. Paper-white flowers with purple stripes on the labellum; named after the propagator's wife (Fig. 98).

'Beesiana White'. The ground colour is a pale yellow-green (1D — green/yellow group), and the streaking on the labellum is barely discernable. The name is, strictly, a misnomer.

Roscoea cautleyoides

'Blackthorn Strain'. This was raised from seedlings of 'Kew Beauty' by Robin White of Blackthorn Nursery, Hampshire. The flowers are variable, in shades of white and purple, or white with purple striations (Fig. 99).

'Dark Beauty'. A very dark-coloured form (N79A — purple group), found as a seedling by Dr Spearing in 1963.

'Deep violet form'. Plant 27 cm tall; flowers a brilliant violet (83A — violet group); labellum 3 cm long. Chance seedling.

'Dwarf yellow form'. Plant 12 cm tall; flowers deep greenish yellow (1A — green/yellow group); labellum 3 cm long. Holds its colour well. Chance seedling.

'Early Purple'. 24 cm tall; smaller flowers than 'Purple Giant' and a duller purple (N77B — purple group). Very short staminodes.

160 THE GENUS ROSCOEA
HYBRIDS AND CULTIVARS

Fig. 102 (left). ***Roscoea cautleyoides* 'Purple Giant'** cultivated and photographed by Roland and Gay Bream, Cruckmeole.
Fig. 103 (right). ***Roscoea cautleyoides* 'Kew Beauty'** cultivated at Kew and photographed by Richard Wilford, R.B.G., Kew.

Fig. 104. ***Roscoea cautleyoides* 'Reinier'** cultivated and photographed by Roland and Gay Bream, Cruckmeole.

'Grandiflora' (see 'Kew Beauty')

'Jeffrey Thomas'. Similar to the species, but the labellum is somewhat paler (155C — white group) than the rest of the flower. Named after the propagator's son (Fig. 100).

'Kew Beauty'. (syn. 'Grandiflora'). Larger flowers than the species, but somewhat paler (ID); 50 cm tall; a very fine plant (Fig. 103).

'Off-white form'. Plant 27 cm tall; flowers off-white with pale purple striations; labellum 3 cm long. Chance seedling.

'Old Purple'. (CLDX 687). Bluish tinge to the leaves; to 18 cm tall, with fine purple flowers. (N81A — purple/violet group)

'Purple Giant'. Flowers large, deep purple (N79B — purple group), c. 30 cm tall (Fig. 102).

'Reinier'. Large yellow (1C — green/yellow group) flower, infused with a hint of green. About 20 cm tall, and earlier to flower than the species (Fig. 104).

'Vien Beauty'. A smaller version of 'Kew Beauty', around 25 cm. tall (Fig. 106).

'Wine Red'. A seedling with striking wine-red flowers (N79B — purple group)

'Yeti'. 21 cm tall; flowers a deepish yellow (1A — green/yellow group); holds its flowers well and does not fade (Fig. 107).

Fig. 105. *Roscoea purpurea* **'Peacock Eye'** (left), **'Brown Peacock'** (middle) and **'Peacock'** (right) cultivated and photographed by Roland and Gay Bream, Cruckmeole.

THE GENUS ROSCOEA
HYBRIDS AND CULTIVARS

Fig. 106 (left). ***Roscoea cautleyoides* 'Vien Beauty'** cultivated and photographed by Roland and Gay Bream, Cruckmeole.
Fig. 107 (right). ***Roscoea cautleyoides* 'Yeti'** cultivated and photographed by Roland and Gay Bream, Cruckmeole.

Fig. 108. ***Roscoea humeana* 'Tall White'** cultivated and photographed by Roland and Gay Bream, Cruckmeole.

Roscoea humeana

'Purple Streaker'. Dwarf; only 13 cm tall. *Purpurea*-like flowers; white labellum heavily streaked with purple (N82A — purple/violet group). Could be a hybrid.

'Rosemoor Plum'. Dark plum colour (N79B — purple group).

'Striped form'. 15 cm tall, with green-yellow flowers (1D — green/yellow group); labellum striped purple (83B — violet group). Chance seedling.

'Tall White' — see 'White form tall' (Fig. 108).

'Violet-purple form'. 25 cm tall with impressive bright flowers (N81A — purple/violet group).

'White form dwarf'. 15 cm tall, with smaller flowers than the next.

'White form tall'. An impressive plant 20 cm tall with pure white flowers. Found in the Lijiang Plain, Yunnan (Fig. 108).

Roscoea purpurea

Many of the cultivars of *Roscoea purpurea* have been propagated by Rene Zijerveld, and have been separated out from seedlings according to height and the presence or absence of red pigment in stem and leaves.

'Brown Peacock' — red stem; red leaves (Fig. 105).

'Gigantea'. From seed collected by Chris Chadwell (CC 1757). The plant resembles *Roscoea auriculata*, and this could well be what it really is. The flower is somewhat darker (N81C — purple/violet group) and has a well-divided labellum and white staminodes.

'Nico'. Green-stemmed and some 40 cm tall. The flowers are larger than the Peacock Strain, with a labellum 5 cm wide (Fig. 111).

'Niedrig'. Small; only 25 cm tall. The flower is slightly darker with a hint of red (N81C — purple/violet group) and the labellum is fairly narrow (3 cm wide). It can have either a green or red stem (Fig. 109).

The Peacock Strain is vigorous (35–55 cm tall), and is so named because by manipulation the flower can be made to resemble a peacock. The flower is a pale mauve (N82D — purple group). The labellum is some 4 cm wide.

'Peacock' — green stem; green leaves (Fig. 105).

'Peacock Eye' — red stem; green leaves (Fig. 105).

Fig. 109. ***Roscoea purpurea* 'Niedrig'** cultivated and photographed by Roland and Gay Bream, Cruckmeole.

'Red Gurkha'. Collected in Nepal in August 1992. The plant has a beautiful brick-red flower (47A — red group) and there are forms with both red and green stems. Labellum 4 cm long. BBMS 45 — the red-stemmed form (Plate 3); 24 cm tall. BBMS 47 — the green stemmed form; 35 cm tall.

'Vincent' is green-stemmed and some 40 cm tall. The flowers are slightly darker (N81C — purple/violet group) and fairly narrow; the labellum is only 3 cm wide. There is a tendency for red pigment to appear on the underside of the mature leaves (Fig. 111).

'Wisley Amethyst'. Collected in Assam in June 1938 by Kingdon Ward (KW 13755). It has been growing at Wisley for the last 60 years and was formerly known as "*purpurea alba*". This was one of 200 plants selected to celebrate 200 years of the Royal Horticultural Society. The plant is 38 cm high, the flower is white, and the labellum is heavily stained with violet (83B — violet group). There are two prominent white lines along the centre of each lobe of the labellum, which is 5 cm long and 3 cm wide. It flowers early, in June. (Figs. 26, 112)

Roscoea scillifolia

'longifolia'. This is identical to *Roscoea scillifolia* 'Pink Form' and was originally supplied by the late Jack Drake (Inshriach Alpine Nursery). Note that *R. longifolia* Baker is a synonym of *R. alpina* Royle.

Roscoea tumjensis

'Purple Giant'. Marketed by Rene Zijenveld, but as far as I can see it is identical with the typical form.

'Sino-purpurea'. This too seems to be identical with *Roscoea tumjensis*. Note that *R. sinopurpurea* Stapf is a synonym of *R. cautleyoides* Gagnep. forma *sinopurpurea* (Stapf) Cowley.

Fig. 110. **Roscoea purpurea** cultivars cultivated commercially at Hillegom, Holland.

Suppliers of *Roscoea* species, hybrids and cultivars

Blackthorn Nursery (Robin White), Kilmeston, Alresford, Hampshire, SO24 0NL

Crûg Farm Plants (Bleddyn & Sue Wynn-Jones), Griffith's Crossing, Caernarfon, Gwynedd, LL55 1TU

Dix & Zijerveld (Rene Zijerveld), Postbus 142, Zandlaan 4, 2180AC Hillegom, Holland (Wholesale only). (Fig. 110).

Europa Nurseries, P.O. Box 17589, London, E1 4YN

Hartside Nursery (Neil Huntley), Hartside Nursery Garden, Alston, Cumbria, CA9 3BL

Hill View Hardy Plants, Worfield, Bridgnorth, Shropshire, WV15 5NT

Long Acre Plants (Nigel & Michelle Rowland), Charlton Musbrove, Somerset, BA9 8EX

Mary Green, The Walled Garden, Hornby, Lancaster, LA2 8LD

Paul Christian, P.O. Box 468, Wrexham, LL13 9XR

Potterton & Martin (Bob Potterton), Moortown Road, Nettleton, Caistor, Lincolnshire, LN7 6HX (species only)

Note
In choosing your plants it is important to check that, if of wild origin, they have been obtained legally. It is also important that they have been cultivated on a basis which is sustainable longterm.

10. CULTIVATION

By Richard Wilford

Roscoeas look best when planted in the garden, where their flowers can peer through low-growing perennials or from beneath shrubs. Different species will prefer different situations but in general they can be grown on a rock garden, raised bed, in a herbaceous border or woodland garden, and will gradually increase to form dense colonies. However, there is a good reason for growing them in pots under glass, as many growers do, and that is to allow control over the amount of moisture they receive, especially in winter when they are dormant. Too much water at this time of year and the fleshy roots will rot and the plants will be lost. In a climate such as Britain's, where rain occurs all year round, this is an important consideration.

When planting roscoeas in the garden, it is also important to consider how much light they will receive. Partial shade is ideal, as too much sun in summer will cause the leaves to curl up and the plant will suffer. Flowering may also be affected in hot weather, as the blooms will not open fully before shrivelling in the heat. On the other hand, in deep shade the plants will become etiolated and weak.

Growing roscoeas under glass means that water is controlled, but the plants will need good ventilation and shading in summer, and the pots must be protected from freezing solid in winter. Choosing the right soil mix is important and, because the root run is restricted, frequent watering will be required when the plants are in growth.

Ideally, roscoeas should be planted outside, but with a backup collection under glass to cover for any losses that may occur. If conditions are favourable, roscoeas can live for many years in a garden, with minimal attention. They are still considered unusual garden plants, but as they gradually become better known they will surely be grown more widely.

ROSCOEAS IN THE OPEN GARDEN

The first *Roscoea* to flower in the garden is usually the yellow form of *R. cautleyoides*, which appears in late April in southern England. This is soon followed by the yellow and purple forms of *R. humeana*. The main flowering season for roscoeas is June to August but if the weather is favourable, — not too hot and dry — they can still be flowering in September or even October. *R. purpurea* is often the last to emerge, pushing through the soil in June and flowering from July to September. If the summer is cool and wet, some of the early flowering species can have a second burst of growth and produce new flowers in autumn, particularly *R. cautleyoides*. So, if you grow a selection of species, you can have roscoeas in flower from April to October.

The most important factor to consider when planting roscoeas outside is their position. The perfect planting position is one where there is plenty of sunlight for the shoots but where the roots are shaded, at least for part of the day, to keep them cool. The soil should be moist throughout the summer but never waterlogged in winter; in other words, it should be moisture retentive but well-drained. Reasonable drainage is essential because the plants can be watered in summer but if they are sitting in wet, heavy clay all winter, they are likely to rot away.

A rock garden can provide the all conditions required for growing roscoeas successfully. Because it is raised above ground level, the soil is well-drained, and the rocks can provide shade to the roots while allowing the shoots to grow into the light. Planting close to rocks allows the roots of the plant to grow under the rock, where the soil retains moisture even during dry spells. This is important for roscoeas, as they do not want to dry out in summer. Planting against east or west facing rocks is best, as this ensures the plants are not in full sun all day.

Some *Roscoea* species are more tolerant of less than ideal conditions than others. *R. cautleyoides* can grow well in a fairly hot, sunny position, whereas the smaller *R. alpina* really does require light shade for a significant part of the day. *R. purpurea*, *R. humeana* and *R. auriculata* are also fairly tolerant of dryness in summer, although the flowers of the first can suffer in the heat, particularly in the red-flowered *R. purpurea* 'Red Gurkha'. In a hot, sunny position, these flowers do not emerge fully from the upper leaf sheaths but with the onset of cooler weather, the flowers revert to their normal position, hanging over the top of the pseudostem.

Roscoea scillifolia is one of the most vigorous species in the garden, especially the pale pink-flowered form. This forms large, dense clumps that seed over a wide area and flower throughout the summer months, from late June to the end of August. The dark-flowered form tends to flower earlier, in May, and is more upright in appearance, but it too will gradually spread to cover a wide area.

Fig. 111. **Roscoea purpurea 'Vincent'** (left) and **'Nico'** (right) with *Potentilla*, cultivated and photographed by Roland and Gay Bream, Cruckmeole.

Fig. 112. **_Roscoea purpurea_** forma **_alba_** with _Geranium_ 'Spinners' cultivated and photographed by John Fielding, Sheen.

Species that require a little more care when choosing a planting position include _Roscoea alpina_, _R. capitata_, _R. forrestii_, _R. tumjensis_ and _R. wardii_. For these, make sure the soil does not dry out completely in summer and provide some shade for part of the day. _R. schneideriana_ seems to be the hardest to please in the open garden and resents any dryness in summer. Careful positioning near rocks, to provide protection from the heat of the sun, is necessary to keep this species alive on a rock garden.

Roscoeas appear to be fully hardy if planted deep enough, although not all the species have been subjected to open garden cultivation to allow a thorough assessment of hardiness in the genus. Several of the more commonly grown species have been successfully cultivated in US zone 5 so in Britain there should really be no problem.

The dormant _Roscoea_ plant should be planted at least 10 cm deep in the soil, to protect it from freezing. Frost should not penetrate this deep in Britain but in areas that are subjected to much colder temperatures, deeper planting may be necessary. A thick bark mulch will also provide some frost protection. A reliable freezing winter, with snow cover, is actually advantageous, as it ensures the rhizomes are kept dry. In areas where temperatures fluctuate above and below freezing and winter precipitation is heavy it is harder, but not impossible, to grow a range of roscoeas in the open. The fact that most, if not all, _Roscoea_ species can be grown outside in Britain indicates how tolerant this genus is of conditions different to those experienced in its natural habitat.

When planting roscoeas it may be necessary to amend the soil by adding grit and sand to improve drainage although, as previously mentioned, the soil on a rock garden should already be free draining. Addition of organic matter can also improve drainage, especially with clay soil. Roscoeas have no particular preference for acid or alkaline soil, so a pH somewhere around neutral is best. Some gritty sand placed around the rhizome, where the fleshy roots join, will help prevent rotting at this point. Once the roscoea has been planted in its hole, the soil is replaced and firmed. It is then well watered to make sure the soil has settled around the roots.

Do not panic if it seems a long time before growth appears above ground. It may be early summer before there is any sign of life. If the right position has been chosen there is little to do after planting. Feed can be applied in spring, in the form of a sprinkling of blood, fish and bone or bone meal, forked into the soil. In a few years the clumps may become congested and the plants will need lifting and dividing. Otherwise they can be left to their own devices, apart from tidying away the old leaves in autumn.

Like a rock garden, a raised bed will have naturally free draining soil but unless the bed is in the shadow of a building or high wall, it will be fully exposed to the sun. This may be ideal for alpine plants and bulbs but most roscoeas will suffer unless frequently watered in summer. Low-growing shrubs can provide some shade and having the plants raised up means that the smaller species, like *Roscoea tibetica* and *R. schneideriana* are more easily appreciated.

A woodland garden tends to be too shaded for most roscoeas, although *Roscoea alpina* seems to like such conditions, as long as the shade is not too deep. The edge of a woodland garden, where there is only partial shade, is more suitable but the humus-rich, moist soil may be too wet in winter for some species.

Fig. 113. **Roscoea cautleyoides** with *Corydalis flexuosa* cultivated and photographed by John Fielding, Sheen.

Fig. 114. ***Roscoea cautleyoides*** with *Hosta* 'August Moon' cultivated and photographed by John Fielding, Sheen.

Most gardeners will want to plant roscoeas in a mixed herbaceous border and this is a perfectly acceptable way to grow them. With the right companion plants they can do very well and their unusual blooms provide a real talking point in the garden. However, it is worth remembering that some species continue to grow after flowering and their leaves can take up a significant amount of space. *Roscoea humeana* has the widest leaves but they are only just emerging when the flowers are open. By late summer the clumps of foliage my reach 40 or 50 cm across.

Heavy, sticky soil will need to be modified to improve drainage in winter and only the taller, more vigorous roscoeas such as *Roscoea auriculata*, *R.* 'Beesiana', *R. cautleyoides*, *R. humeana*, *R.* 'Kew Beauty' and *R. purpurea,* will do well in a mixed border. Plant them with low-growing perennials that will not cause too much shade or compete for moisture in the soil. Hardy geraniums are good partners, such as *Geranium sanguineum* var. *striatum* or *G. wallichianum*. The early-flowering *G. clarkei* 'Kashmir White' will accompany early roscoeas and the later, sprawling *G.* 'Anne Folkard' will flower with *R. purpurea.*

On a rock garden, sedums make good partners, growing beneath the roscoeas, and the Himalayan *Androsace lanuginosa*, which requires the same growing conditions, will trail its silvery stems around the roscoeas' shoots. Examples from the rock garden at Kew include *Roscoea humeana* flowering with the purple-blue *Aquilegia pubiflora* and the yellow *R. cautleyoides* growing through *Scutellaria alpina*. All these plants allow the taller roscoeas to flower above or alongside them. Smaller species, such as *R. tibetica* and *R. schneideriana* will be overshadowed so they need to be grown in a more open situation, with less competition for light or water. For these species, pot cultivation is often more suitable.

GROWING ROSCOEAS UNDER GLASS

Growing roscoeas under glass gives complete control to the grower over the amount of water the plants receive. The full range of species can be grown in pots kept in a cool glasshouse or cold frame but this method does present its own problems. Plants grown under glass are in danger of overheating on a hot day but good ventilation and shading can help prevent this.

A glasshouse for growing roscoeas will need vents in the roof, will preferably have a door at each end to allow air to pass along the length of the house, and will have openings, either windows or louvres, at plant height along the sides. This ventilation, combined with shading, will go a long way to reducing summer temperatures. Nevertheless, the glasshouse will still get very warm at times and the plants will need frequent watering. The *Roscoea* collection at Kew was grown in a glasshouse for several years but now it has been moved to a cold frame, and the plants look much healthier, are less stressed in summer and more vigorous.

The advantage of a cold frame is that the glass frame lights can be completely removed in summer, when the roscoeas are in growth. This allows maximum ventilation and the frame can still be shaded if necessary. In autumn and winter the frame lights are replaced to keep the plants dry and provide some frost protection.

The pots in a cold frame should be plunged in sand up to their rim, to prevent them freezing solid, which would kill the roots. Plunging the pots in moist sand will also keep them cooler in summer, and the soil will not dry out so quickly. If clay pots are used, moisture can pass from the sand, through the sides of the pot, to the soil. As well as helping to keep the soil moist when the

Fig. 115. ***Roscoea humeana*** with *Geranium* cultivated at Cruckmeole and photographed in June 2000 by Richard Wilford, R.B.G., Kew.

Fig. 116. Divided roots of **Roscoea cautleyoides** in November 1999, cultivated and photographed by Richard Wilford, R.B.G., Kew.

roscoeas are in full growth, this can be a useful way to water the plants just as they are coming into growth or dying down, when you may not want to apply water directly to the pot.

If you are unable to plunge your pots then you will have to provide some frost protection, such as a frost-free glasshouse. However, as the plants are dormant in winter, you do not have to provide light so any cool but frost-free space is suitable. Root growth can start quite early in the year so the pots will need to be watered and brought into the open in late winter.

Roscoeas should be repotted every year, in late winter, to provide fresh soil and allow the health of the plants to be checked. If they are doing well, they may outgrow their pot so a larger one is needed or the clump may need dividing. This is done by carefully pulling the growing points apart and untangling the fleshy roots without damaging or detaching them. They are then replanted in a new, clean pot.

The pot should be deep enough to accommodate the long fleshy roots. The growing point needs to be between 5 and 10 cm below the surface and the fleshy roots can often reach 15 cm long or more so a shallow pot or pan is not suitable. As a guide, plant three flowering size plants in a 15 cm diameter pot. Use a gritty, loam-based soil mix with some coir (coconut fibre) or leaf-mould added, to help retain moisture. Place a few handfuls of soil in the bottom of the pot. Arrange the fleshy roots of the *Roscoea* and trickle soil around them, holding them in place until the soil mix supports them. Stop just before the soil reaches the rhizome and place some gritty sand around the growing point. This is where rot often sets in, so extra drainage is beneficial. Then continue adding soil mix until the level is about 1 or 2 cm from the pot rim. The last layer should be a mulch of grit. This protects the soil surface during watering and helps retain moisture and keep down weeds.

After repotting, the plants should be watered thoroughly, until water can be seen running out of the pot's drainage holes. This ensures the soil is completely wet. Allow the soil to dry out before watering again. It is much better to water this way than to give a dribble of water every day or two, as over-watering is less likely to occur. Once growth appears above the surface, increase watering and throughout the summer, never allow the soil to dry out. A low nitrogen, high potash, slow release fertilizer can be incorporated into the soil mix to feed the plants. Alternatively, a liquid feed can be applied when watering. Stop feeding and reduce watering in autumn, as the plants die down, and in winter, if you use clay pots, only keep the plunge sand moist.

PROPAGATION

As explained above, roscoeas can be divided when they are repotted in late winter. Care must be taken not to break the fleshy roots, as this can weaken the plant. If division is not carried out, the plants will become more and more congested and pot-bound, resulting in a loss of vigour. The same is true in the garden but even if the plants look healthy, dividing them allows you to spread them around or give some away (Fig. 116).

Roscoeas in the garden are best divided in late autumn, as the dying leaves will help in locating the roots. Lift the plants, cut away any remaining leaves and wash off the soil. Carefully ease the growing points apart until you have a number of individual plants in a range of sizes. It is surprising how many plants you can get from one established clump. They can then be replanted around 10 to 20 cm apart, depending on the size of the species.

The alternative way to increase your stock is to grow roscoeas from seed. Some species freely seed around in the garden with no human intervention. *Roscoea scillifolia* is very good at this. If you want some control over what is grown and where, then you should collect the seed yourself.

Species that have an exserted peduncle will hold their seed pods well above the leaves. The peduncle may elongate after flowering, as in *Roscoea cautleyoides*, which holds its seed pods 40 or 50 cm above ground. Species with a short peduncle will hold their seed pods within the leaf sheaths. For example, in *R. alpina* the seed pod is at ground level and can be reached by easing the leaves apart. Other species, such as *R. tibetica* and *R. purpurea*, have their seed pods in the upper leaf sheaths and the seeds have to be picked out from the top of the pseudostem.

Roscoea seed is best sown fresh, as soon as it has been collected in autumn. Older seed will not germinate as quickly or evenly as fresh seed, and may not germinate at all, but soaking old seed for 24 hours can improve the germination rate. Sow the small brown seeds in a pot of free-draining soil mix and cover them with a thin layer of soil and a layer of fine grit to protect the soil surface. Water the pot by standing it in a tray of water so that the soil is soaked by capillary action without disturbing the seeds. Once moisture can be seen at the top of the pot, you know all the soil is wet. Label the pot with the plant name and date of sowing.

Place the seed pots outside, so they are exposed to the cold in winter. Do not let the pot dry out. Germination usually takes place in early spring, as the weather warms up, and at this point the seed pots should be given some shelter, preferably in a glasshouse to protect the delicate seedlings. Once they are large enough to handle, the seedlings should be pricked out into individual pots. If left in their seed pot too long, the roots can become tangled and the fragile seedlings may be damaged as they are separated.

Seedling roscoeas will need regular potting on as they grow and flowering will generally take place after three or four years, although it may happen after only two years. If a collection of species is grown it is important to bear in mind that hybridisation can occur. Garden hybrids like *Roscoea* 'Beesiana' and *R.* 'Kew Beauty' have arisen this way but plants raised from seed should be checked to see if they are true to the parent species. It is worth growing them on to flowering size before planting them out in the garden so see if they have come true. Hybrids are only rarely an improvement on the species. It is also worth remembering that seed offered in seed exchanges may well be the result of hybridisation.

PESTS AND DISEASES

Fortunately, roscoeas are fairly resilient plants. If they are healthy and well-grown they rarely succumb to pests. Aphids will attack new growth, especially the inflorescence, and may cause some damage, and in the garden slugs and snails may munch on the new shoots. There are many ways to deter slugs and snails, some more effective than others, but on a rock garden, a mulch of sharp grit will certainly put them off.

The enclosed space of a glasshouse, with its warmer temperatures, is a breeding ground for a number of pests, including aphids and red spider mite, which often attack roscoeas. Red spider mite, in particular, revels in hot dry conditions, so moving your plant collection into a cold frame, with its cooler air, will help enormously with this problem. In a glasshouse, both aphids and red spider mite can be treated by biological control. The predators *Phytoseiulus* and *Amblyseius* are both effective against red spider mite and there are a number of predators of aphids, including *Aphidius* and *Aphidoletes*. These beneficial insects are little use in an open cold frame but in a glasshouse they can be very effective if introduced early in the year, before the pest populations are out of control.

In a garden or cold frame, aphids can be sprayed with a horticultural soap solution. This drowns the aphids by breaking the surface tension of the water which then fills their breathing pores. Chemical controls are also very effective, especially those containing imidacloprid, which can be applied to pots as a soil drench.

The main cause of death in roscoeas is rotting of the roots in soil that is too wet. Improving soil drainage, increasing ventilation and careful watering will all help avoid this problem. Losses can also occur if the plants are left to dry out in summer. If they are planted in the garden, they should be moved to a more suitable location. Well grown plants should live for many years with few problems.

LIST OF EXSICCATAE

The herbarium specimens seen for this study are arranged alphabetically according to collector and numerically for each collector. The species number in the text is given in bold in ( ) brackets and the Botanical Institutions where they may be found in [ ] brackets.

The Botanical Institutions are coded as follows:
Arnold Arboretum [A], Breslau (Wroclaw) [WRSL], Brussels [BR], Calcutta [CAL], Canton (Guangzhou) [IBSC], Cambridge University (UK)[CGE], Dublin [TCD], Edinburgh [E], Hong Kong [HKU], Kathmandu [KATH], Kew [K], Kunming [KUN], Smith Herbarium [LINN], Liverpool [LIV], Michigan University [MICH], North India [?BSHC], Paris [P], St. Petersburg [LE], Smithsonian [US], Vienna [WU].

Expedition abbreviations:
ACEX = Alpine Garden Society Expedition to China (1994).
AGSES = Alpine Garden Society Expedition to Sikkim (1983).
CLDX = Chungtien-Lijiang-Dali Expedition (1990).
KEKE = Kew-Edinburgh-Kathmandu Expedition to NE Nepal (1989).
SABE = Sino-American Botanical Expedition to Yunnan (1984).
SBLE = Sino-British Expedition to Lijiang Co., W. Yunnan, China (1987)

List of species: **1** alpina; **2** purpurea; **3** capitata; **4** ganeshensis; **5** nepalensis; **6** tumjensis; **7** auriculata; **8** bhutanica; **9** brandisii; **10** australis; **11** wardii; **12** schneideriana; **13** scillifolia; **14** cautleyoides; **15** humeana; **16** forrestii; **17** tibetica; **18** debilis; **19** praecox; **20** kunmingensis

ACEX 2 (**17**)[K], 19 (**19**)[K], 38 (**19**)[K], 38A (**19**)[K], 236 (**17**)[K], 250 (**15**)[K], 251 (**17**)[K], 344 (**17**)[K], 346 (**17**)[K], 353 (**17**)[K], 406a-e (**17**)[K], 484 (**17**)[K], 981 (**14**)[K]; *AGSES* 319 (**7**)[K]; *Aitchison* s.n. (**1**)[K].

Bailey's collectors s.n. (**2**)[BM], s.n. (**3**)[BM]; *Baker, Miller & Burkitt* 15 (**6**)[K], 16 (**7**)[K], 109, 110, 112 (**2**)[K]; *Baker, Burkitt, Miller & Shrestha* 2, 3 (**1**)[K], 13 (**3**)[K], 34, s.n. (**4**)[K], 40, 41 43, 45, 46, 47(**2**)[K], 55, 56 (**3**)[K]; *Barclay & Synge* 2348, 2416, 2702 (**1**)[K]; *Bedi* 144, 928, 1276 (**8**)[K]; *Benham* s.n. (**1**)[BM]; *Billiet & Leonard* 6495 (**2**)[BELG, K]; *Bithika* s.n. (**2**)[G]; *Blinkworth* s.n. (**2**)[K]; *Bowes Lyon* 3224 (**8**)[BM], 15739 (**2**)[BM]; *Brandis* s.n. (**9**)[CAL, K]; *Buchanan* s.n. (**2**)[K, LINN, LIV].

Cave 111 (**7**)[K]; *Cavalerie* 4763 (**19**)[K]; *Ching* 20321, 20433 (**14**)[KUN], 20443 (**15**)[KUN], 20759 (**12**)[KUN], 20827, 20828 (**17**)[KUN], 20986 (**17**)[KUN], 21421 (**12**)[KUN], 30133 (**15**)[KUN]; *Clarke* 17590A (**9**)[CAL], 28272 (**1**)[K], 38420 (**9**)[K], 38491C (**9**)[K], 38584A, 44268A (**9**)[G], 44607A (**9**)[K], 46355A (**7**)[K]; *CLDX* 282 (**17**)[K], 483 (**17**)[K], 686 (**12**)[K],

687, 772 (**14**)[K], 773 (**12**)[K], 1209 (**17**)[K], 1523 (**17**)[K], 1543 (**17**)[K], s.n. (**16**)[K]; *Codrington* 236, 238 (**2**)[BM]; *Collett* ?156 (**1**)[K], 633 (**2**)[K]; *Cooper* 378 (**7**)[E], 1300 (**8**)[BM, E], 2526, 3252 (**8**)[BM], 6009 (**10**)[E]; *Cribb* 41 (**15**)[K]; *Cult. RBG Edinburgh* s.n. (**15**)[E].

Dalhousie s.n. (**1**)[G, K], s.n. (**2**)[E, G, K]; *Davies* s.n. (**2**)[K]; *Deane* s.n. (**1**)[K]; *Delavay* 92, 231 (**14**)[P], 2659 (**14**)[K, P], 2685, 2685 bis (**13**)[P], 3283 (**13**)[CAL, K, P], 4491 (**14**)[K, P], s.n. (**17**)[P]; *Dorji, Pearce & Cribb* 48 (**2**)[K], 89 (**8**)[K]; *Drummond* 1944, 20898, 20949 (**1**)[K], 22734 (**1**)[G, K], 22735, 22736, 23185 (**1**)[K], 26413 (**2**)[K], 26414 (**2**)[G, K], 26412, 26416 (**1**)[K]; *Ducloux* 401 (**19**)[K], 597, 601 (**19**)[P], 688 (**18**)[P], 768 (**18**)[K], 1257 (**18**)[E], 5203 (**18**)[K, P], s.n. (**19**)[K, P]; *Dun* 85 (**10**)[CAL]; *Dungboo* 9 (**7**)[CAL, US], 54 (**7**)[CAL, K], 56 (**8**)[CAL, K], 58 (**1**)[CAL, K], 59 (**7**)[CAL, K], 4244 (**8**)[CAL, K], s.n. (**8**)[CAL, K]; *Duthie* 559 (**1**)[G, K], 7365 (**1**)[K], 24973 (29) (**1**)[K], 24985 (**2**)[K]; *Dhwoj* 04 (**6**)[BM, E], 490 (**1**)[BM, E].

Edgeworth 9 (**1**)[K], 10, s.n., s.n. (**2**)[K]; *Elwes* s.n. (**1**)[G].

Falconer 1233 (**2**)[G]; *Farrer* 1766 (**11**)[E]; *Feng* 872 (**14**)[KUN], 1037, 1262 (**15**)[KUN], 1388 (**17**)[KUN], 1972 (**12**)[KUN]; *Fielding* s.n. (**1**)[K], s.n. (**2**)[K]; *Forrest* 2070 (**14**)[E, K], 2178 (**14**)[E, K], 2396 (**17**)[BM], 2687 (**14**)[HKU, K], 4806, 4807, 4808 (**17**)[BM, E], 4809 (**14**)[BM, E], 5890 (**14**)[BM, E, IBSC, K], 5930 (**15**)[K], 5969 (**14**)[K], 5988 (**17**)[K], 6354 (**13**)[K], 6387 (**14**)[K], 6401 (**12**)[BM, E, IBSC, K], 6513 (**13**)[BM, E, IBSC, K], 6917 (**18**)[E, K], 7041 (**17**)[K], 7050 (**14**)[E, K], 8456 (**18**)[E, K], 10218 (**15**)[K], 10638 (**17**)[BM, K], 10655 (**12**)[K], 10657 (**13**)[BM, E, K], 10945 (**12**)[K], 11726 (**16**)[BM, E, K], 12910 (**12**)[E], 19236 (**17**)[K], 21437 (**15**)[CAL, E, K, US], 21447 (**15**)[E, K, US], 21527 (**15**)[E, K, US], 23229 (**17**)[E, K, US].

Gamble 4420A, 4445A, 4484A, 4485A, 4493A, 4493C, 4585A (**1**)[K], 4653A, 4663A, 4663C (**2**)[K], 4663E (**2**)[CAL], 25819, 26988 (**1**)[K], 26999 (**2**)[K]; *Gardner* 525 (**6**)[BM], 790 (**6**)[BM, E], 847 (**3**)[BM]; *Gebauer* s.n. (**16**)[WU]; *Gould* 251, 356, 912, 925 (**8**)[K], 2937 (**1**)[K]; *Gregory & Gregory* s.n. (**12**)[BM], s.n. (**19**)[BM]; *Grey-Wilson & Phillips* 149 (**1**)[K], 274 (**2**)[K]; *Grierson & Long* 116, 1826 (**8**)[E]; *Griffith* 5736, s.n. (**9**)[K].

Haines 2301 (**1**)[K]; *Halliwell* 34 (**3**)[K], 35, 198 (**2**)[K]; *Hancock* 170 (**19**)[K]; *Handel-Mazzetti* 2253 (**14**)[WU], 2491 (**15**)[WU], 2966 (**17**)[WU], 3166 (**13**)[WU], 3351 (**17**)[WU], 4145 (**15**)[K, WU], 4152 (**12**)[WU], 4153 (**17**)[WU], 4154 (**15**)[WU], 9510 (**11**)[WU]; *Hara et al.* 6183 (**1**)[BM], 723602 (**2**)[BM]; *Harsukh* s.n. (**1**)[?BSHC, K]; *Henry* 11102, 11102A (**18**)[K], 11102B (**18**)[E, K], 11102C (**18**)[K], 11117 (**19**)[E, K]; *Hingston* 148 (**1**)[K], 165 (**1**)[K]; *Hobson* s.n. (**1**)[K]; *Hooker* s.n. (**2**)[K], s.n. (**7**)[G, K, TCD], 1452, s.n. (**9**)[K], s.n. (**9**)[G, TCD]; *Howell* 44 (**18**)[E, K], 333 (**18**)[E]; *Huggins* 2 (**2**)[BM].

Jacquemont 1024 (**1**)[K, LIV, P], 2410 (**2**)[K, P].

Kachkarov s.n. (**17**)[LE]; *Kanai, Hara & Ohba* 721776, 723600 (**3**)[BM], 723607 (**2**)[BM]; *KEKE* 258 (**2**)[E, K, KATH], 291 (**7**)[E, K]; *King* s.n. (**1**)[CAL, K], s.n. (**2**)[CAL, K]; *King's Collector* 10 (**7**)[G], 53, 55 (**7**)[K], 57 (**1**)[K], 60, 61 (**7**)[CAL, K], 62 (**1**)[K], 63 (**7**)[CAL, K], 454 (**8**)[CAL, K]; *Kingdon Ward* 1639, 3199 (**17**)[E], 4104 (**16**)[E], 4355 (**14**)[E], 4376 (**15**)[E], 6885, 7112, 8382 (**11**)[K], 9682 (**11**)[BM], 10476 (**11**)[BM], 11529, 13755 (**2**)[BM], 18682 (**9**)[BM], 19623 (**11**)[BM], 22124 (**10**)[BM, E, K], 22292, 22380 (**10**)[BM]; *Koelz* 4961 (**1**)[G], 33255 (**9**)[K, MICH].

Legge s.n. (**1**)[K]; *Lester-Garland* s.n.(?1349) (**1**)[K]; *Limpricht* 855 (**18**)[WRSL]; *Lobb* 62 (**9**)[K]; *Long, McBeath, Noltie & Watson* 131 (**7**)[E]; *Long & Noltie* 115 (**7**)[E]; *Ludlow & Sherriff* 50 (**8**)[BM], 309 (**2**)[BM], 2275, 3123 (**8**)[BM]; *Ludlow, Sherriff & Hicks* 16377 (**8**)[BM, E], 16439 (**1**)[BM], 18911 (**8**)[BM], 20845 (**2**)[BM, E]; *Luo* 20 (**1**)[K], 37(**17**)[K].

Madden s.n. (**2**)[K]; *Maire* 267 (**12**)[E], 467 (**19**)[E, K], s.n. *Bonati* Sér.B 2989, 3462 (**12**)[US]; *Mann* 347 (**9**)[CAL]; *McCosh* 65 (**6**)[BM]; *McLaren* B67 (**17**)[BM, E, K], B105, B106 (**16**)[BM, E, K], B128 (**14**)[K], L63 (**12**)[BM, K], V47A (**19**)[E]; *Miehe* (4)80 (**5**)[BM]; *Munro* 2151 (**1**)[K].

Norton 319 (**7**)[K].

Polunin 691 (**3**)[BM], 1949 (**2**)[BM]; *Polunin, Sykes & Williams* 362 (**5**)[BM, E, K], 436 (**2**)[BM], 2266, 2460 (**1**)[BM], 2500 (**2**)[BM], 4340 (**1**)[BM, E, G], 4381 (**5**)[BM], 4391 (**5**)[BM, E, K]; *Pratt* 518 (**17**)[BM, K].

Ribu & Rhomoo 5520 (**7**)[CAL, E, K]; *Rich* 247 (**1**)[K], 321 (**2**)[K]; *Rock* 3475, 3793 (**15**)[E, US], 4448 (**13**)[E, K, KUN, US], 4549 (**15**)[E, US], 4589 (**17**)[K, US], 4617 (**17**)[KUN], 4709 (**17**)[K, KUN, US], 4726 (**12**)[E, K, KUN, US], 4759 (**13**)[E, US], 4888 (**12**)[E, US], 5069 (**14**)[K, US], 5486 (**17**)[E, US], 16008 (**15**)[E, US], 16009 (**15**)[E, KUN, US], 17811 (**12**)[E, US], 23852 (**15**)[B, BM, E, G, IBSC, K, US], 24831 (**14**)[BM, E, G, K, KUN, US], 24930 (**14**)[BM, E, K, US], 25147 (**17**)[G, K]; *Royle* s.n. (**1**)[TCD], s.n. (**2**)[K]

SABE 998 (**14**)[KUN]; *SBLE* 542 (**14**)[K], 612 (**15**)[K]; *Schilling* 418 (**1**)[K, KATH], 609, 1185 (**2**)[K]; *Schilling, Sayers et al.* 418 (**1**)[K]; *Schneider* 1192 (**15**)[A, E, G, K, US], 1200 (**14**)[A, G, K], 1232 (**14**)[E, G, K, US], 1625 (**17**)[A, G, K], 1770 (**12**)[A, G, K, US], 1971 (**17**)[A, G, K], 2070 (**17**)[A, G, K], 2264 (**12**)[G, US], 3795 (**14**)[A, G, K]; *Schoch* 179 (**19**)[G, K, WU]; *Scully* C66 (**2**)[CAL, K], s.n. (**3**)[K]; *Sherriff* 7312 (**1**)[BM]; *SICH* 1027 (**15**)[K]; *Smith* s.n. (**1**)[K], s.n. (**2**)[K]; *Spencer Chapman* 152 (**7**)[K], 298 (**1**)[K]; *Stainton* 1067 (**7**)[BM], 1198 (**2**)[BM], 3833 (**3**)[BM], 4594 (**6**)[BM], 7174 (**7**)[BM]; *Stainton, Sykes & Williams* 958 (**1**)[G], 1596 (**2**)[BM, CAL, E], 1628 (**5**)[BM, E], 2836 (**1**)[G], 3021 (**1**)[BM], 3328 (**5**)[BM, E]; *Stewart* 13363 (**1**)[G]; *Stonor* 83 (**1**)[K]; *Strachey & Winterbottom* 1(**2**)[K], 2 (**1**)[K], 1986 (**2**)[K].

Tessier-Yandell 280 (**9**)[K]; *Thomson* 1259 (**1**)[K], 1341 (**2**)[K], s.n., s.n. (**1**)[G, K, TCD], s.n. (**2**)[G, TCD]; *Tong* (**16**)[KUN]; *Tsai* 52960, 52961 (**12**)[KUN], 58141 (**17**)[KUN].

Ujfalvy s.n. (**1**)[P].

Venning 10 (**10**)[K].

Wallich 6528 (**2**)[K, TCD], 6528A (**2**)[G], 6528B, 6528E (**2**)[G, K], 6528D (**1**)[G], 6529 (**3**)[BM, CGE, E, G, K, TCD]; *Wang* 63787, 64075, 64462 (**17**)[KUN], 70672 (**14**)[KUN], 71341 (**12**)[KUN]; *Williams* 967 (**7**)[BM]; *Wilson* 4601 (**17**)[K]; *Wollaston* 281 (**7**)[K]; *Wyss-Dunant* 1079 (**7**)[G], 1192 (**2**)[G].

Younghusband s.n. (**7**)[BM, G, K]; *Yü* 5445 (**15**)[KUN], 15229, 16596 (**17**)[KUN], 16863, 17925 (**12**)[KUN], 19717 (**11**)[E].

Zang 8921 (**14**)[KUN]; *Zimmermann* 752 (**7**)[G], 769 (**1**)[G], 826 (**1**)[G], 1082A (**2**)[G], 1229, 1824A (**3**)[G].

BIBLIOGRAPHY & REFERENCES

Anderson, E. B. (1956). Autumn Bulbs for Garden and Cold House. *Journal of the Royal Horticultural Society* 81: 125–126.

Anon. & authors various. *Journal of the Royal Horticultural Society*.
(1910–1911) 36(II): 400. (1912) 38(2): 192. (1913–1914) 39: cxxxiii–iv. (1915) 41(2): 206. (1916–1917) 42(1): 44; 42: clxi. (1920–1921) 46: xlii; 46: lxiii. (1923) 48(2 & 3): 203, 208. (1924) 49(2): 152. (1927) 52: lvi. (1929) 54(2): 458. (1935) 60(11): 485, 489. (1936) 61(12): cxxxiii, clv. (1938) 63(8): 353; 63(9): 409. (1939) 64(4): 178; 64(9): 394. (1940) 65(8): 231; 65(9): 271. (1942) 67(11): 371. (1944) 69(6): 151. (1945) 70(5): 124; 70(6): 155. (1946) 71(6): 152, lxii. (1947) 72(3): 123. (1948) 73(5): 133; 73(12): lxxx. (1949) 74(5): 206. (1950) 75(4): 154; 75(5): 174; 75(6): 229–230; 75(8): 316. (1951) 76(6): 182. (1952) 77(6): 197. (1954) 79(8) 356. (1957) 82(2): 89. (1958) 83(3):136. (1959) 84(5): 206–212, 84(6): 255. (1960) 85(5): 221; 85(6): 250. (1962) 87(6): 249. (1963) 88(2): 94. (1964) 89(4): 173; 89(5): 173. (1965) 90(8): 347. (1967) 92(1): 15; 92(6): 247. (1969) 94(5): 225, 227, 230. (1973) 98(3): 113. (1981) 106(12): 510. (1982) 107(2):49. (1986) 111(8): 375. (1987) 112(6): 272–273.

Anon. & authors various. *Gardener's Chronicle*.
(1841) I: 149, 719. (1889) VI: 186, 249. (1890) VIII: 190, 221, 251, 278. (1891) 9: 54. (1900) 28: 203. (1916) 59: 300. (1920) 67: 101. (1928) 84: 12, 29, 57. (1930) 88: 294. (1938) 104: 47. (1939) 105: 2, 3; 106: 266. (1940) 108: 6. (1944) 116: 34. (1946) 120: 23, 26. (1947) 122: 99, 101. (1948) 124: 143. (1953) 133: 21, 23. (1954) 136: 174, 176. (1955) 137: 125, 127. (1956) 139: 226, 432. (1958) 143: 371; 144: 9. (1960) 147: 355. (1961) 150: 209, 473. (1963) 153: 96; 154: 242. (1964) 156: 10–11, 529. (1970) 167(15): 31, (20): 23; 168(10): 35. (1977) 181(17): 21–22. (1980) 187(7): 31.

Anon. & authors various. Bulletin, Department of Medicinal Plants, Kathmandu 6 (1976). *Flora of Langtang*: 216; *Index Florae Yunnanensis*: (1984). T2: 1912–1913; *Flora Sichuanica* 10: 595 (?1994); *Flora Yunnanica* 8: 572 (1997).

Bailey, L.H. (1916). *Standard Cyclopedia of Horticulture* 3(P–Z): 2999. The Macmillan Company, New York.

Baker, J.G. (1894). Roscoea. In: J. D. Hooker (ed.). *Flora of British India* 6: 207–208. L. Reeve, London.

Baker, W.J. (1994). Three men and an orchid. *Bulletin of the Alpine Garden Society* 62(1): 101–114; 62(2): 189–193.

Balfour, B. & Smith, W.W. (1916). *Roscoea humeana*. Notes from the Royal Botanic Garden Edinburgh 9, 42: 122.

Batalin, A.T. (1895). *Roscoea tibetica*. Trudy Imperatorskago S. Peterburgskago Botaniceskago Sada 14, 8: 183.

Bloom, A. (1991). *Alan Bloom's Hardy Perennials*: 140. Batsford, London.

Branney, T. (2002). More than a hint of ginger. *The Garden* 127(9): 718–723.

Brightman, C. (1991). Some Interesting Plants at the Shows 1989–1990. *Bulletin of the Alpine Garden Society* 59(1): 55.

Bryan, J.E. (1989). *Roscoea-Zingiberaceae. Bulbs*: 319–320. Christopher Helm, London.

Burtt, B.L. (1972). General introduction to papers on *Zingiberaceae*. *Notes from the Royal Botanic Garden Edinburgh* 31(2): 155–165.

—— & Smith, R.M. (1972a). Tentative keys to the subfamilies, tribes and genera of *Zingiberaceae*. *Notes from the Royal Botanic Garden Edinburgh* 31(2): 171–176.

—— (1972b). Key species in the taxonomic history of *Zingiberaceae*. *Notes from the Royal Botanic Garden Edinburgh* 31(2): 177–227.

Cameron, M.L. (1998). Yunnan — People, Places and Plants. *Rhododendron* 38: 20–28.

Chadwell, C. *et al*. (1991–92). *Himalayan Plant Association Newsletter* — issues 2–4.

Chandler, G. (1953). *William Roscoe of Liverpool 1753–1831*. B.T. Batsford Ltd.

Chittenden, F.J. (ed.) (1951). *The RHS Dictionary of Gardening* 4: 1822. Clarendon Press, Oxford.

Chowdhery, H.J. & Wadhwa, B.M. (1984). *Flora of India* series 2. *Flora of the Himachal Pradesh* 3: 695. Botanical Survey of India, Howrah.

Clay, S. (1937*). The Present-Day Rock Garden*: 546. T.C. & E.C. Jack, Ltd., London & Edinburgh.

Collett, H. (1902). *Flora Simlensis: XCIX (Scitamineae)*: 509–510. W. Thacker & Co., London.

Cowan, J.M. (1938). A Review of the genus *Roscoea. The New Flora & Silva* 11(1): 17–28.

—— (1952). The Journeys + *Roscoea. The Journeys and Plant Introductions of George Forrest VMH*: 33, 220–221. Published for the Royal Horticultural Society of London by Oxford University Press.

Cowley, E.J. (1980). A new species of *Roscoea (Zingiberaceae)* from Nepal. *Kew Bulletin* 34(4): 811–812.

—— (1982). A revision of *Roscoea (Zingiberaceae). Kew Bulletin* 36(4): 747–777.

—— (1984). *Roscoea. The European Garden Flora* (II): 125. Cambridge University Press.

—— (1994). *Roscoea schneideriana. The Kew Magazine* 11(1): 13–18.

—— (1996). AGS China Expedition (ACE). The Plants — *Roscoea. Bulletin of the Alpine Garden Society* 64(2): 239–242.

—— (1997). *Roscoea praecox. Botanical Magazine* 14(1): 2–6.

—— (1997). *Roscoea alpina. Botanical Magazine* 14(2): 77–81.

—— & Baker, W. (1994). *Roscoea purpurea* 'Red Gurkha'. *The Kew Magazine* 11(3): 104–109.

—— & —— (1996). *Roscoea ganeshensis. Botanical Magazine* 13(1): 8–13.

—— & Wilford, R. (1998). *Roscoea tumjensis. Botanical Magazine* 15(4): 220–225.

—— & —— (1998). *Roscoea capitata. Botanical Magazine* 15(4): 226–230.

—— & —— (2000). *Roscoea humeana* forma *lutea. Botanical Magazine* 17(1): 22–28.

Cox, E.H.M. (1926). *Farrer's Last Journey*. Dulau & Co., Ltd., London.

—— (ed.) (1930). *The Plant introductions of Reginald Farrer*. New Flora and Silva Ltd., London.

Cullen, J. (1973). William Roscoe's Monandrian Plants of the order *Scitamineae. Notes Royal Botanic Garden, Edinburgh* 32(3): 417–421.

Dang, R. (1993). *Flowers of the Western Himalayas*: 115, 118. Indus Publishing, New Delhi.

Dierl, W. (1968). Zur Nahrungsaufname von *Corizoneura longirostris. Ergebnisse der Forsch. Untern. Nepal-Himalaya* 3: 70–81.

Elliott, Joe (1986). Forty years a nurseryman. *Bulletin of the Alpine Garden Society* 54(2): 119, 122.

Elliott, Jack (1997). *The Smaller Perennials*: 146–147. Batsford, London.

—— (1998). Show Reports 1997. *Bulletin of the Alpine Garden Society* 66(1): 66.

Everett, T.H. (1982). *The New York Botanical Garden Illustrated Encyclopedia of Horticulture* (9): 2980. Garland Publishing, New York.

Farrer, R. (1925). *The English Rock Garden* (II): 224–225. T.C. & E.C. Jack, London.

Fletcher, T.B. & Son, S.K. (1931). A veterinary entomology for India, XIV. *Indian Journal of Veterinary Science & Animal Husbandry* 1: 192–199.

Gagnepain, M.F. (1901, published in 1902). Zingibéracées nouvelles. *Bulletin de la Société Botanique de France* 48: LXXIII–LXXVII.

Gentil, L. (1907). *Roscoea sikkimensis. Liste des plantes cultivées dans les serres chaudes et coloniales du jardin botanique de l'État à Bruxelles*: 169. P. Weissenbruch, Brussels.

Grey, C.H. (1938). *Roscoea. Hardy Bulbs* 2: 346–357. Williams & Norgate Ltd., London.

Grey-Wilson, C. (1988). Journey to the Jade Dragon Snow Mountains, Yunnan 1. *Bulletin of the Alpine Garden Society* 56(1): 24–28.

—— (1993). Alpine Jewels from the Jade Dragon Mountains. *The Garden* 118(3): 102–103.

—— (1996). Beyond Big Snow Mountain. *The Garden* 121(6): 335–339.

Guan, Kaiyun (ed.) *et al.* (1998). *Highland Flowers of Yunnan*. Yunnan Science & Technology Press.

Hajra, P.K. & Verma, D.M. (eds.) (1996). *Flora of Sikkim* 1: 132. Botanical Survey of India, Calcutta.

Hall, A. (1995). Plant Awards 1994–1995. *Bulletin of the Alpine Garden Society* 63(4): 384–386.

—— (2000). Plant Awards 1999–2000. *Bulletin of the Alpine Garden Society* 68(4): 503.

Handel-Mazzetti, H. (1927). A Botanical Pioneer in South West China. *Österreichischer Bundesverlag*. English translation by David Winstanley, (1996): 23, 24, 36, 63, 67. Published by David Winstanley, Brentwood.

—— (1936). *Symbolae Sinicae VII* Anthophyta: 1320–1321. J. Springer, Vienna.

Hara, H. (1966). *The Flora of the Eastern Himalaya*: 423. University of Tokyo Press.

—— *et al.* (1963). *Indo-Japanese Botanical Expedition to Sikkim and Darjeeling, 1960. Spring flora of Sikkim Himalaya*: 121. Hoikusha, Osaka.

——, Stearn, W.T. & Williams, L.H.J. (1978). *An Enumeration of the Flowering Plants of Nepal* 1: 61. British Museum of Natural History, London.

Harvey, F.W. (ed.) (1914). *The Garden* 78: 159.

Haw, S.G. (1992). Ecology. The Flora of China — an introduction. *Sino-Himalayan Plant Association Newsletter* — issue 5.

Hilton, E. (1982). Some moisture loving plants. *Bulletin of the Alpine Garden Society* 50(3): 221.

Holttum, R.E. (1974a). A Commentary on Comparative Morphology in *Zingiberaceae. The Gardens Bulletin, Singapore* 27(2): 155–165.

—— (1974b). *Zingiberales. Encyclopaedia Britannica*: 1150–1155.

Hooker, J.D. (1852). *Roscoea purpurea. Botanical Magazine* 78: t. 4630.

Hooker, W.J. (1825). *Roscoea purpurea. Exotic Flora* 2: t. 144. W. Blackwood, Edinburgh.

Horaninow, P. (1862). *Prodromus Monographiae Scitaminearum*: 20–21. Typus Academiae Caesareae Scientiarum, Petropoli.

Huxley, A. (ed.) (1992). *The New RHS Dictionary of Gardening* (4): 134–135. Macmillan Press, London.

Jones, A. (1983). Roscoeas and the L.B.G. *Bulletin of the Alpine Garden Society* 51(1): 70, 84–86.

Karthikeyan, S., Jain, S.K., Nayar, M.P. & Sanjappa, M. (1989). *Flora of India* series 4. *Florae Indicae Enumeratio: Monocotyledonae*: 297. Botanical Survey of India, Calcutta.

Kingdon Ward, F. (1923). The Flora of the Tibetan Marches. *Journal of the Royal Horticultural Society* 48: 201–212.

—— (1923). *Mystery Rivers of Tibet*. Seeley, Service & Co., Ltd., London.
—— (1924a). The Flora of the upper Irrawaddy. *Journal of the Royal Horticultural Society* 49: 148–156.
—— (1924b). *From China to Hkamti Long*: 42, 312. E. Arnold & Co., London.
—— (1924c). *The Romance of Plant hunting*: 111–112, 115, 126. E. Arnold & Co., London.
—— (1926). *Riddle of the Tsangpo Gorges*. E. Arnold & Co., London.
—— (1926–27). The Sino-Himalayan Flora. *Proceedings of the Linnean Society of London*: 67–74.
—— (1934). Explorations in Tibet 1933. *Proceedings of the Linnean Society of London* 1933–34(pt. III): 110–113.
—— (1937). *Plant Hunter's Paradise*. J. Cape, London.
—— (1946). Additional Notes on the Botany of north Burma. *Journal of the Bombay Natural History Society* 40(2): 380–389.
—— (1952). Plant hunting in Assam. *Journal of the Royal Horticultural Society* LXXVII(6): 205–214.
—— (1953). The Assam Earthquake of 1950. *The Geographical Journal* CXIX(2): 169–182.
—— (1954). Report on the Forests of the North Triangle, Kachin State, North Burma. *Journal of the Bombay Natural History Society* 52 (2 & 3): 304–320.
—— (1955). Plant-Hunting in the Triangle, North Burma. *Journal of the Royal Horticultural Society*. (April): 174–190.
—— (1957). The Great Forest Belt of North Burma. *Proceedings of the Linnean Society of London* 1955–56 (pts 1&2): 87–96.
—— (1959). A Sketch of the Flora and Vegetation of Mount Victoria in Burma. *Acta Horti Gothoburgensis* XXII: 53–58, 66–74.
—— (1960). *Pilgrimage for Plants*. G.G. Harrap, London.
Kitamura, S. (1955). Zingiberaceae. *Fauna & Flora of Nepal Himalaya* 1: 98–99. Fauna and Flora Research Society, Kyoto University.
Kumar, S. (2001). *Zingiberaceae of Sikkim*. Deep Publications, New Delhi.
Lancaster, R. (1987). *Garden Plants for Connoisseurs*: 159–163. Unwin Hyman, London.
—— (1989). *Travels in China, A Plantsman's Paradise*: 265, 271. Antique Collectors' Club, Woodbridge, Suffolk.
—— (1992). Plants that should be better known. *The Garden* 117(8): 376–378.
—— (1995). *A Plantsman in Nepal*: 33, 37. Antique Collectors' Club, Woodbridge, Suffolk.
Lang, D.C. (1991). A Visit to the Lhonak Valley of Sikkim. *Bulletin of the Alpine Garden Society* 59(3): 269.
Lemaire, Ch. (1852). *Le Jardin Fleuriste*. Gand.
Léveillé, H. (1917). *Catalogue des Plantes du Yunnan*. Scitaminacées: 260. Le Mans.
Liang, Y. (1988). Pollen morphology of the family *Zingiberaceae* in China — Pollen types and their significance in the Taxonomy. *Acta Phytotaxonomica Sinica* 26(4): 265–281.
Lindley, J. (1840). Roscoea purpurea. *Botanical Register* 26: t. 61.
Loddiges, G. (1828). Roscoea purpurea. *Botanical Cabinet* t. 1404.
Loesener, Th. (1923). Über einige *Roscoea*-Arten aus Yunnan. *Notizblatt des Botanischen Gartens und Museums zu Berlin-Dahlem* 8: 599–600.
Lynch, R.I. (1882). On a Contrivance for Cross-fertilisation in *Roscoea purpurea* with incidental reference to the structure of *Salvia grahami*. *Journal of the Linnean Society (Botany) London* 19: 204–206.
Lyte, C. (1989). *Frank Kingdon Ward — the last of the great plant hunters*. J. Murray, London.

Mahanty, H.K. (1970). A cytological study of the *Zingiberales* with special reference to their taxonomy. *Cytologia* 35: 13–49.
Malla, S.B. *et al.* (eds.) (1976). Flora of Langtang and Cross Section Vegetation Survey. *Bulletin, Department of Medicinal Plants, Thapathali, Kathmandu, No. 6, HMG Nepal.*
—— (1986). Flora of Kathmandu Valley. *Bulletin, Department of Medicinal Plants, Thapathali, Kathmandu No. 11, HMG Nepal.*
Mathew, B. (1984). Yes! There were bulbs as well. *Bulletin of the Alpine Garden Society* 52(3): 266, 268.
—— (1987). *The Smaller Bulbs*: 149–151. B.T. Batsford, London.
McBeath, R. (1996). Collecting on Big Snow Mountain. *The Garden* 121(7). 410–413.
—— (2001). Oriental Beginnings. *Proceedings of the Seventh International Rock Garden Plant Conference (Alpines 2001)*: 10, 33. Scottish Rock Garden Club and The Alpine Garden Society.
McLean, B. (1997). *A Pioneering Plantsman — A. K. Bulley and the Great Plant Hunters*. The Stationery Office, London.
—— (2004). *George Forrest Plant Hunter*. Antique Collectors' Club, Woodbridge, in association with the Royal Botanic Garden, Edinburgh.
Meikle, R.D. (1973). The Development of Garden Plants from Wild Species. In: P.S. Green (ed.), *Plants: wild and cultivated*: 114–123. Published for the Botanical Society of the British Isles by E.W. Classey, Hampton.
Miau, R. (1995). Plantae Rockianae. *Acta Scientiarum Naturalium Universitatis Sunyatseni* 34(3): 81–82.
Morley, B. (1980). Augustine Henry. *The Garden* 105(7): 285–289.
Naithani, H.B. (1990). *Flowering plants of India, Nepal and Bhutan*. Surya Publications, Dehra Dun.
Needham, E. (2002). Two Nepalese Roscoeas. *The Alpine Gardener* 70(2): 152–153.
Ngamriabsakul, C. (2001). *The Systematics of the Hedychieae (Zingiberaceae), with emphasis on Roscoea Sm.* University of Edinburgh PhD thesis (not published).
—— & Newman, M.F. (2000). A New Species of *Roscoea* Sm. (*Zingiberaceae*) from Bhutan and Southern Tibet. *Edinburgh Journal of Botany* 57(2): 271–278.
——, —— & Cronk, Q.C.B. (2000). Phylogeny and disjuction in *Roscoea* (*Zingiberaceae*). *Edinburgh Journal of Botany* 57(1): 39–61
Nicholson, G. (ed.) (1886). *The Illustrated Dictionary of Gardening* (III): 326. L.U. Gill, London.
Noltie, H.J. (1994). *Roscoea. Flora of Bhutan* 3(1): 194–196. Royal Botanic Garden, Edinburgh.
Nordhagen, R. (1932). Zur Morphologie und Verbreitungsbiologie der Gattung *Roscoea* Sm. *Bergens Museums Arbok 1932 Naturvidenskapelig rekke* nr. 4.
Ohba, H. & Ikeda, H. (eds.) (1999). *A Contribution to the Flora of Ganesh Himal, Central Nepal*. University Museum, University of Tokyo.
Polunin, O. & Stainton, J.D.A. (1997). *Flowers of the Himalaya*: 408 & Plate 119. Oxford University Press.
Rao, V.S. (1963). The epigynous glands of *Zingiberaceae*. *New Phytologist* 62: 342–349.
Rehder, A. (1930). Ernest Henry Wilson. *Journal of the Arnold Arboretum* XI: 180–192.
Ridley, H.N. (1916). On endemism and the mutation theory. *Annals of Botany* XXX No. CXX: 551–574.
Robertson, C. (1897). Seed crests & myrmecophilous dissemination in certain plants. *Botanical Gazette* XXIII: 288–289.
Rolfe, R. (1998). Plant Awards 1997–1998. *Bulletin of the Alpine Garden Society* 66(4): 473–474.
—— (2000). Plant Awards 1998–1999. *Bulletin of the Alpine Garden Society* 68(2): 272–273.

—— (2002). Plant Awards 2001–2002. *The Alpine Gardener* 70(4): 446, 447, 490.
Roscoe, H. (1833). *The Life of William Roscoe* (2 vols.). T. Cadell, Strand & W. Blackwood, Edinburgh.
Roscoe, W. (1807). A new arrangement of the plants of the Monandrian class usually called Scitamineae. *Transactions of the Linnean Society of London* 8: 330–357.
—— (1816). *Transactions of the Linnean Society of London* 11: 270–282.
—— (1828). *Roscoea purpurea*. *Monandrian Plants of the order Scitamineae* sect. 2: t. 64. G. Smith, Liverpool.
Rose, D. (1981). Dr. Nathaniel Wallich. *The Garden* 106(5): 196–209.
Roxburgh, W. (1832). *Flora Indica*: 476. W. Thacker, Serampore.
Royle, J.F. (1839). *Roscoea alpina*. In: *Illustrations of the botany and other branches of the natural history of the Himalayan mountains, and of the Flora of Cashmere*: 361, t. 89. W.H. Allen & Co., London.
Salzen, H. (2000). *The Rock Garden* 26(4): 321–323.
Schilling, A. (1969). *Journal of the Royal Horticultural Society* 94: 230.
—— (1985). Frank Kingdon Ward (1885–1958). *The Garden* 110(6): 264–269.
Schumann, K. (1904). *Zingiberaceae*. Engler's *Das Pflanzenreich* IV, 46: 115–122. W. Engelmann, Leipzig.
Smith, J.E. (1805–1808). *Roscoea purpurea*. *Exotic Botany* 2: 97, t. 108. R. Taylor & Co., London.
—— (1822). *Roscoea capitata*. *Transactions of the Linnean Society of London* 13: 461.
Smith, R.M. (1981). *Synoptic keys to the genera of the Zingiberaceae pro parte: Zingibereae, Globbeae, Hedychieae & Alpineae (in part)*: 8–12. Royal Botanic Garden, Edinburgh.
Spearing, J.K. (1977). A note on closed leaf-sheaths in *Zingiberaceae-Zingiberoideae*. *Notes from the Royal Botanic Garden, Edinburgh* 35(2): 217–220.
Stansfield, H. (1955). William Roscoe, Botanist. *Liverpool Libraries, Museums & Arts Committee Bulletin* 5, 1–2: 18–61
Stapf, O. (1925). *Roscoea humeana*. *Botanical Magazine* 151: t. 9075.
—— (1926). *Roscoea cautleoides, R. sinopurpurea*. *Botanical Magazine* 151: t. 9084.
Stearn, W.T. (1992). *Botanical Latin* 4th edition. David & Charles, Newton Abbot, Devon.
Strachey, R. (1906). *Catalogue of the Plants of Kumaon*: 184. L. Reeve & Co., London.
Sutton, S.B. (1974). *In China's Border Provinces. The turbulent career of Joseph Rock, botanist-explorer.* Hastings House, New York.
Sykes, W. (1956). 1954 Expedition to Nepal: part 2. *Journal of the Royal Horticultural Society* 81: 6–14.
Taylor, M. & H. (1999). New Flowers in the NW Himalaya. *The Rock Garden* 25(4): 376.
Thompson, W. (1890). *Roscoea purpurea* var. *sikkimensis*. *Gardener's Chronicle* ser. 3, 8: 251.
Thomson, A.E. (1982). Ramblings, 4 — with a promise! *Bulletin of the Alpine Garden Society* 50(3): 197.
Tomlinson, P.B. (1956). Studies in the systematic anatomy of the *Zingiberaceae*. *Botanical Journal of the Linnean Society* 55: 547–592.
—— (1962). Phylogeny of the *Scitamineae* — Morphological and Anatomical Considerations. *Evolution* XVI(2): 192–213.
—— (1969). *Anatomy of the monocotyledons 3: Commelinales–Zingiberales*: 341–359. Clarendon Press, Oxford.
Tong, S.Q. (1997). *Zingiberaceae*. *Flora Yunnanica* 8. Science Press, Beijing.
—— (1998). Revision and additional notes of *Zingiberaceae* of Yunnan, China. *Bulletin of Botanical Research, Harbin* 18(2): 137–143.

Troll, W. (1929). *Roscoea purpurea* Sm. eine *Zingiberaceae* mit Hebelmechanismus in den Blüten. *Planta*. Bd. 7.

Upward, M. (1984). The Sikkim Adventure. *Bulletin of the Alpine Garden Society* 52(3): 212.

Wallich, N. (1832a). *Roscoea gigantea*. *Plantae Asiaticae Rariores* 3: 22. Treuttel & Wurtz, London, Paris, Strasburg.

—— (1832b). *Roscoea procera*. *Plantae Asiaticae Rariores* 3: t. 242. Treuttel & Wurtz, London, Paris, Strasburg.

—— (1832c). *Roscoea capitata*. *Plantae Asiaticae Rariores* 3: 35, t. 255. Treuttel & Wurtz, London, Paris, Strasburg.

West, J.P. & Cowley, J. (1993). Floristic notes & chromosome numbers of some Chinese *Roscoea* (*Zingiberaceae*). *Kew Bulletin* 48(4): 799–803.

Whitehead, J. (ed.) (1990). *Himalayan Enchantment. Frank Kingdon Ward, an anthology*. Serindia, London.

Wight, R. (1853). *Icones Plantarum Indiae Orientalis*: VI, 17 and t. 2013. J.B. Pharoah, Madras.

Wilford, R. (1998). Rejuvenation at Kew. *Bulletin of the Alpine Garden Society* 66(2): 178.

—— (1999). Roscoeas for the Rock Garden. *Bulletin of the Alpine Garden Society* 67(1): 93–101.

—— (2000). Roscoeas. *Gardens Illustrated* 50: 62–69.

Wilson, E.H. (1913). *A Naturalist in Western China*. Methuen & Co., Ltd., London.

Wright, C.H. (1903). Forbes & Hemsley — Index Florae Sinensis 3 — Monocots. from *Journal of the Linnean Society* (*Botany*) 36: 67.

Wu Delin (1981). *Roscoea*. *Flora Reipublicae Popularis Sinicae* 16(2): 48–55. Science Press, China.

—— et al. (eds.) (1996). *Proceedings of the 2nd Symposium on the family Zingiberaceae*. Zhongshan University Press, Guangzhou, China.

—— & Larsen, K. (2000). *Flora of China* 24: 362–366. Science Press, Beijing & Missouri Botanical Garden Press, St. Louis.

Yi, Guo & Wu, Qigen (1991). The comparative anatomy of the vegetative organs in *Roscoea* Smith (*Zingiberaceae*). *Acta Botanica Austro Sinica. Guangzhou* 7: 39–53, 167–168.

Zhu, Z.Y. (1988). *Acta Phytotaxonomica Sinica* 26(4): 315–316.

INDEX OF SCIENTIFIC NAMES

Accepted names in **bold**, synonyms in *italics*, hybrid and cultivar names 'roman'. Page numbers for illustrations are in **bold**.

Roscoea Sm. 18
   alpina Royle 39, **42**, 44
      f. **alpina** 40
      f. **pallida** Cowley 40, **43**, 45
      var. *minor* 43
   alpina sensu auct. hort. 105
   alpina sensu Hara 67
   auriculata K. Schum. 22, 73, **74**, **76**, **77**, 78
      'Floriade' 154, 159
      'Pink Form' 159
      'White Cap' 154
   auriculata sensu Hara 52
   australis Cowley 86, **87**, 89
   ×**beesiana** 153, 154, **155**, 156, **157**, 158
      'Beesiana Gestreept' 159
      'Beesiana Monique' 159
      'Beesiana White' 159
      'Monique' 154, **155**
      'White Form' 154
   bhutanica Ngamriab. 79, **80**, **81**, 82
   blanda K. Schum. 93, 144
      var. *limprichtii* Loes. 145
      var. *pumila* Hand.-Mazz. 93
   brandisii (Baker) K. Schum. 83, **84**, 86
   capitata Sm. 55, **57**, **58**, 59
      var. *purpurata* Gagnep.
      var. *scillifolia* Gagnep. 105
   cautleyoides Gagnep. 21, 22, 107, 115, 170, 171, 173
      var. **cautleyoides** 110
         f. **atropurpurea** Cowley 117
         f. **cautleyoides** 108, 109, 112, 115, 116, 118

         f. **sinopurpurea** (Stapf) Cowley 110, 111, 112, 113, 114, 115, 117
      var. **pubescens** (Z.Y. Zhu) T.L. Wu 117
      var. *purpurea* 'Dark Beauty' 117
      'August Beauty' 110, 154
      'Bees' Dwarf' 154
      'Beesii' 154
      'Blackthorn Strain' 154, **155**, 159
      'Dark Beauty' 154, 159
      'Deep violet form' 159
      'Dwarf yellow form' 159
      'Early Purple' 154, 159
      'Grandiflora' 154, 161
      'Himalaya' 154
      'Jeffrey Thomas' 154, **156**, 161
      'Kew Beauty' 109, 154, **160**, 161
      'Off-white form' 161
      'Old Purple' 161
      'Purple Giant' 154, **160**, 161
      'Reinier' 154, **160**, 161
      'Vien Beauty' 161, **162**
      'Yeti' 161, **162**
   chamaeleon Gagnep. 115
   debilis Gagnep. 142, **143**, 144
      var. **debilis** 145
      var. **limprichtii** (Loes.) Cowley 145
   ?exilis Horan. 43
   exilis sensu K. Schum. 52, 85
   forrestii Cowley 22, 131, 1330
      f. **forrestii** 132, **133**, **134**, 135
      f. **purpurea** Cowley 135
   ganeshensis Cowley & W.J. Baker 60, **61**, 63, **64**, 65

humeana Balf. f. & W.W. Sm. **22**, 118, 128, 172
    f. **alba** Cowley **122**, 129
    f. **humeana** 119, 120, 121, 122, 123, 124, 128, 129
    f. **lutea** Cowley 123, 124, 125, 126, 127, 130
    f. **tyria** Cowley **130**
    'Inkling' 130, 154
    'Purple Streaker' 163
    'Rosemoor Plum' 154, 163
    'Striped form' 163
    'Tall White' **162**
    'Violet-purple form' 162
    'White form dwarf' 162
    'White form tall' see 'Tall White'
intermedia Gagnep. 43
    var. *anomala* Gagnep. 150
    var. *macrorhiza* Gagnep. 150
    var. *minuta* Gagnep. 139
    var. *plurifolia* Loes. 139
kunmingensis S.Q. Tong 151, **152**
    var. **elongatobractea** S.Q. Tong 152
    var. **kunmingensis** 152
longifolia Baker 43
nepalensis Cowley 66, **67**
praecox K. Schum. 20, 145, **146**, 148, 149, 150
procera Wall. 52
pubescens Z.Y. Zhu 117
purpurea Sm. **8**, **22**, 45, 48, 110
    f. **alba** Cowley **55**, 169
    f. **purpurea** 46, 49, 50, 51
    f. **rubra** Cowley **53, 54**, 55
    var. *auriculata* (K. Schum.) Hara 78
    var. *exilis* (Horan) Baker 43
    var. *exilis sensu* Baker 52
    var. *gigantea* Wall. 52
    var. *pallida* Hort. 52
    var. *procera* (Wall.) Baker 52
    var. *purpurea* Hara 52

    'Brown Peacock' 154, **161**, 163
    'Gigantea' 163
    'Nico' 154, 163, **168**
    'Niedrig' **163**
    'Peacock' 154, **161**, 163
    'Peacock Eye' 154, **161**, 163
    'Red Cap' 154
    'Red Gurkha' **54**, 164
    'Short Form' 154
    'Tall Form' 154
    'Vincent' 154, 164, **168**
    'Wisley Amethyst' 164
purpurea sensu Hooker 85
purpurea sensu Royle 43
schneideriana (Loes.) Cowley **19**, **21**, 96, 97, 98, 99, 100
scillifolia (Gagnep.) Cowley **22**, 101, **104**, 105
    f. **atropurpurea** Cowley **102**, **106**
    f. **scillifolia** 102, 105
    'longifolia' 164
sichuanensis R.H. Miau 130
sikkimensis Hort. ex Gentil 78
sinopurpurea Stapf 115
tibetica Batalin **22**, 135, **139**, 140, 141
    f. **alba** Cowley **136**, 140, **141**
    f. **albo-purpurea** Cowley **136**, 141
    f. **atropurpurea** Cowley 141, **142**
    f. **rosea** Cowley **136**, 142
    f. **roseo-purpurea** Cowley **136**, 141
    var. *emarginata* S.Q. Tong 139
tumjensis Cowley 68, 69, **70**, 71, 72
    'Purple Giant' 164
    'Purple King' 154
    'Sino-purpurea' 164
wardii Cowley **22**, 91, **92**, 93, **94**, 95
yunnanensis Loes. 100, 115
    var. *dielsiana* Loes. 100
    var. *purpurata* (Gagnep.) Loes. 115
    var. *schneideriana* Loes. 100
    var. *scillifolia* (Gagnep.) Loes. 105